Advances in Solid Oxide Fuel Cells VII

T0201176

Contents

INTERCONNECTS

NOVEL CELL/STACK DESIGN AND PROCESSING

RELIABILITY/DEGRADATION

Advances in Solid Oxide Fuel Cells VII

A Collection of Papers Presented at the 35th International Conference on Advanced Ceramics and Composites January 23–28, 2011 Daytona Beach, Florida

Edited by
Narottam P. Bansal
Prabhakar Singh

Volume Editors
Sujanto Widjaja
Dileep Singh

The American Ceramic Society

A John Wiley & Sons, Inc., Publication

Published by John Wiley & Sons, Inc., Hoboken, New Jersey.
Published simultaneously in Canada.

For general information on our other products and services or for technical support, please contact our Customer Care Department within the United States at (800) 762-2974, outside the United States at (317) 572-3993 or fax (317) 572-4002.

Wiley also publishes its books in a variety of electronic formats. Some content that appears in print may not be available in electronic formats. For more information about Wiley products, visit our web site at www.wiley.com.

Library of Congress Cataloging-in-Publication Data is available.

ISBN 978-1-118-05989-0

oBook ISBN: 978-1-118-09524-9
ePDF ISBN: 978-1-118-17261-2

ISSN: 0196-6219

Printed in the United States of America.

10 9 8 7 6 5 4 3 2 1

FUEL REFORMING

Preface

The 8th International Symposium on Solid Oxide Fuel Cells (SOFC): Materials, Science, and Technology was held during the 35th International Conference and Exposition on Advanced Ceramics and Composites in Daytona Beach, FL, January 23 to 28, 2011. This symposium provided an international forum for scientists, engineers, and technologists to discuss and exchange state-of-the-art ideas, information, and technology on various aspects of solid oxide fuel cells. A total of 85 papers were presented in the form of oral and poster presentations, including twelve invited lectures, indicating strong interest in the scientifically and technologically important field of solid oxide fuel cells. Authors from fifteen countries (China, Denmark, Germany, Greece, India, Iran, Italy, Japan, Russia, South Korea, Spain, Sweden, Switzerland, United Kingdom and U.S.A.) participated. The speakers represented universities, industries, and government research laboratories.

These proceedings contain contributions on various aspects of solid oxide fuel cells that were discussed at the symposium. Eighteen papers describing the current status of solid oxide fuel cells technology are included in this volume.

The editors wish to extend their gratitude and appreciation to all the authors for their contributions and cooperation, to all the participants and session chairs for their time and efforts, and to all the reviewers for their useful comments and suggestions. We hope that this volume will serve as a valuable reference for the engineers, scientists, researchers and others interested in the materials, science and technology of solid oxide fuel cells.

Narottam P. Bansal, *NASA Glenn Research Center*
Prabhakar Singh, *University of Connecticut*

Introduction

This CESP issue represents papers that were submitted and approved for the proceedings of the 35th International Conference on Advanced Ceramics and Composites (ICACC), held January 23–28, 2011 in Daytona Beach, Florida. ICACC is the most prominent international meeting in the area of advanced structural, functional, and nanoscopic ceramics, composites, and other emerging ceramic materials and technologies. This prestigious conference has been organized by The American Ceramic Society's (ACerS) Engineering Ceramics Division (ECD) since 1977.

The conference was organized into the following symposia and focused sessions:

Symposium 1	Mechanical Behavior and Performance of Ceramics and Composites
Symposium 2	Advanced Ceramic Coatings for Structural, Environmental, and Functional Applications
Symposium 3	8th International Symposium on Solid Oxide Fuel Cells (SOFC): Materials, Science, and Technology
Symposium 4	Armor Ceramics
Symposium 5	Next Generation Bioceramics
Symposium 6	International Symposium on Ceramics for Electric Energy Generation, Storage, and Distribution
Symposium 7	5th International Symposium on Nanostructured Materials and Nanocomposites: Development and Applications
Symposium 8	5th International Symposium on Advanced Processing & Manufacturing Technologies (APMT) for Structural & Multifunctional Materials and Systems
Symposium 9	Porous Ceramics: Novel Developments and Applications

Symposium 10	Thermal Management Materials and Technologies
Symposium 11	Advanced Sensor Technology, Developments and Applications
Symposium 12	Materials for Extreme Environments: Ultrahigh Temperature Ceramics (UHTCs) and Nanolaminated Ternary Carbides and Nitrides (MAX Phases)
Symposium 13	Advanced Ceramics and Composites for Nuclear and Fusion Applications
Symposium 14	Advanced Materials and Technologies for Rechargeable Batteries
Focused Session 1	Geopolymers and other Inorganic Polymers
Focused Session 2	Global Mineral Resources for Strategic and Emerging Technologies
Focused Session 3	Computational Design, Modeling, Simulation and Characterization of Ceramics and Composites
Special Session	Pacific Rim Engineering Ceramics Summit

The conference proceedings are published into 9 issues of the 2011 Ceramic Engineering & Science Proceedings (CESP); Volume 32, Issues 2-10, 2011 as outlined below:

- Mechanical Properties and Performance of Engineering Ceramics and Composites VI, CESP Volume 32, Issue 2 (includes papers from Symposium 1)
- Mechanical Properties and Performance of Engineering Ceramics and Composites VI, CESP Volume 32, Issue 2 (includes papers from Symposium 1)
- Advanced Ceramic Coatings and Materials for Extreme Environments, Volume 32, Issue 3 (includes papers from Symposia 2 and 12)
- Advances in Solid Oxide Fuel Cells VI, CESP Volume 32, Issue 4 (includes papers from Symposium 3)
- Advances in Ceramic Armor VII, CESP Volume 32, Issue 5 (includes papers from Symposium 4)
- Advances in Bioceramics and Porous Ceramics IV, CESP Volume 32, Issue 6 (includes papers from Symposia 5 and 9)
- Nanostructured Materials and Nanotechnology V, CESP Volume 32, Issue 7 (includes papers from Symposium 7)
- Advanced Processing and Manufacturing Technologies for Structural and Multifunctional Materials V, CESP Volume 32, Issue 8 (includes papers from Symposium 8)
- Ceramic Materials for Energy Applications, CESP Volume 32, Issue 9 (includes papers from Symposia 6, 13, and 14)
- Developments in Strategic Materials and Computational Design II, CESP Volume 32, Issue 10 (includes papers from Symposium 10 and 11 and from Focused Sessions 1, and 2)

The organization of the Daytona Beach meeting and the publication of these proceedings were possible thanks to the professional staff of ACerS and the tireless dedication of many ECD members. We would especially like to express our sincere thanks to the symposia organizers, session chairs, presenters and conference attendees, for their efforts and enthusiastic participation in the vibrant and cutting-edge conference.

ACerS and the ECD invite you to attend the 36th International Conference on Advanced Ceramics and Composites (http://www.ceramics.org/daytona2012) January 22–27, 2012 in Daytona Beach, Florida.

SUJANTO WIDJAJA AND DILEEP SINGH
Volume Editors
June 2011

Cell/Stack Development

RECENT DEVELOPMENT OF SOFC CELL AND STACK AT NTT

Reiichi Chiba, Hiroaki Taguchi, Takeshi Komatsu, Himeko Orui, Kazuhiko Nozawa,
Kimitaka Watanabe, Yoshiteru Yoshida, Masayuki Yokoo, Akihiro Miyasaka,
Hajime Arai and Katsuya Hayashi

NTT Energy and Environment Systems Laboratories, NTT Corporation.
3-1, Morinosato-Wakamiya, Atsugi-shi, Kanagawa, 243-0198, JAPAN
E-mail: chiba@aecl.ntt.co.jp

ABSTRACT

We have developed a pinhole free anode support cell (120mm in diameter) with a scandia alumina doped zirconia (SASZ) electrolyte, a $LaNi_{0.4}Fe_{0.6}O_3$ (LNF) cathode and a Ni-SASZ anode. The LNF cathode and SASZ electrolyte were developed by NTT. SASZ is phase transition free electrolyte whose conductivity is three times that of 8YSZ. LNF has several merits including high electrical conductivity, high thermal expansion compatibility with zirconia electrolyte and very high resistance to chromia poisoning. To improve the cathode performance, a composite active layer consisting of LNF and $Ce_{0.9}Gd_{0.1}O_2$ is placed between the electrolyte and the LNF current collection layer. This prevents the LNF from reacting with zirconia ($La_2Zr_2O_7$ forming reaction). The cell was tested at 800°C for 6000h and exhibited good voltage retention (-0.4%/1000h) at $0.4A/cm^2$. We also fabricated a stack consisting of 40 cells and ferritic stainless steel interconnectors containing chromium. The stack operated stably for about 400h at 800°C with humidified CH_4 (S/C=3.0) fuel (U_{fuel}=60%) and air (U_{air}=30%). The maximum efficiency of the stack was 64% (DC, LHV) at U_{fuel}=85%.

INTRODUCTION

Nippon Telegraph and Telephone Corporation (NTT) is a telecommunications company that is a heavy user of electric power accounting for 1% of Japan's commercial electric power consumption. As shown in Fig. 1, the power consumption has been increasing with the increase in telecommunications traffic in Japan [1]. Most of the power is consumed by central offices and data centers, which are located in the centers of cities. The central offices of NTT have been using electric power for the - 48V DC power supply needed for metallic telephone subscriber lines. AC grid power is converted to DC power and supplied to servers in the data centers. Therefore, the solid oxide fuel cell (SOFC) system is a promising alternative way of meeting such electric power demands [2,3]. This is because SOFCs are clean, quiet and efficient DC power generating systems that emit less CO_2 than conventional thermal power generation systems [2,3]. A large amount of the electric power consumed by NTT is used for air conditioning the facilities. If an SOFC system with cogeneration is introduced, this energy demand can also be met. One of our development targets is a medium scale

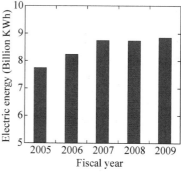

Figure 1. Electric power consumption of NTT Group.

(several kW - several hundred kW) SOFC system for base-load power generation in such facilities, which could help to reduce CO_2 emissions.

We have been investigating anode support planar cells, ferritic stainless steel interconnectors and stacks composed of them. One of the advantages of anode support cells is that it is possible to reduce the thickness of the electrolyte with a wet process such as screen printing [2,4-6]. A thin electrolyte enables the cells to operate at a high current density that leads to high power density generation [2,4-6]. Planar type cells have a simple structure and are suitable for cost reduction [2, 6]. The cost of producing metallic interconnectors is potentially low, because of their very low material costs for high-volume production [2,7,8]. Therefore, SOFC systems using anode support planar cells with stainless steel interconnectors can achieve high performance power generation at a reduced cost. This paper reports NTT's development of SASZ electrolyte material, LNF cathode materials, planar cells using those materials, and stacks employing the cells and stainless steel manifolds.

SASZ FOR ELECTROLYTE MATERIAL

Rare-earth doped zirconia electrolytes are suitable for anode support cells, because their ionic transport number is very close to 1.0 [2,9]. This characteristic is essential for cells with a thin electrolyte, because the amount of electronic leak current density across the electrolyte is proportional to its thickness. Sc_2O_3 doped zirconia (SSZ) has the highest ionic conductivity among the doped zirconias [9], but it has a phase transition around 650°C. As shown in Fig. 2, the ionic conductivity changes by approximately two orders at this temperature. Moreover, the thermal expansion coefficient (TEC) changes abruptly at this transition, as shown in Fig. 3(a) We use Al_2O_3 dopant to suppress this phase transition. In Fig. 4, the phase transition temperature is plotted against the doping amount.

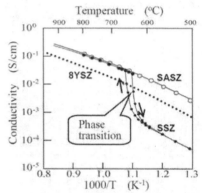

Figure 2. Temperature dependence of ionic conductivity for zirconias, (a) SASZ (doped with 10.5mol% Sc_2O_3 and 0.5mol% Al_2O_3), (b) SSZ (doped with 12mol% Sc_2O_3) and (c) 8YSZ (doped with 8mol% Y_2O_3).

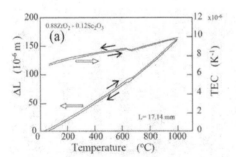
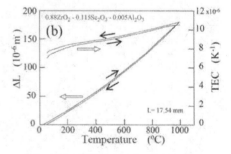

Figure 3. Thermal expansion and thermal expansion coefficient for (a)SSZ(0.88ZrO_2 - 0.12Sc_2O_3), (b)SASZ(0.88ZrO_2-0.115Sc_2O_3-0.005Al_2O_3).

Only a very small amount of Al₂O₃ is required to reduce the phase transition at room temperature (According to the XRD, the sample with X=0.005 was in the cubic phase). Because of this, the ionic conductivity of scandia alumina doped zirconia (SASZ) is almost the same as that of SSZ. The ionic conductivity of SASZ (0.89 ZrO₂ - 0.105 Sc₂O₃ - 0.005 Al₂O₃) at 800°C is 0.1S/cm, which is approximately three times that of conventional zirconia electrolyte (8mol% yttria doped zirconia). As shown in Fig. 3(b), the abrupt TEC change disappeared. These thermal expansion characteristics are very important for practical electrolytes, because the electrolyte layer must withstand thermal cycles caused by the starting and stopping of the operation.

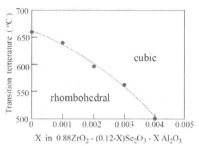

Figure 4. Phase transition temperature and X in 0.88ZrO₂-(0.12-X) Sc₂O₃-X Al₂O₃.

LaNi(Fe)O₃ FOR CATHODE MATERIAL

We have been investigating LaNi(Fe)O₃ as a cathode material [10]. This perovskite type oxide has several advantages including high electronic conductivity, good TEC matching with zirconia, good cathodic performance and high durability against chromium poisoning [10-14].

LaNi(Fe)O₃ PROPERTIES

LaNiO₃ is known to exhibit high electronic conductivity [10]. But it is not stable at the high temperature needed for SOFC operation, and decomposes into NiO and La₂NiO₄ [10]. LaNiO₃ forms a solid solution with LaFeO₃, which is a very stable oxide. This solid solution is stable at or above 800°C when X in LaNi$_{1-X}$Fe$_X$O₃ exceeds 0.3 [10]. In Fig. 5(b), the electrical conductivity at 800°C and the crystal structure at room temperature are plotted against the LaNi$_{1-X}$Fe$_X$O₃ composition. The electronic conductivity was measured in air by the DC four terminal method for well-sintered rod shaped samples. The crystal structures of the synthesized powder samples were analyzed with an X-ray diffractometer at room temperature. The samples were in the rhombohedral phase and the conductivity exhibited the highest value of 580S/cm at 800°C and X=0.4. This value is much higher than that of conventional cathode material (around 200S/cm for La₀.₈Sr₀.₂MnO₃) [2,10]. The thermal expansion coefficient (TEC) for LaNi$_{1-X}$Fe$_X$O₃ is plotted against X in Fig. 5(a). The figure shows the mean values between room temperature and 800°C or 1000°C, respectively. TEC decreased monotonically as X increased. We selected a composition where X=0.4, at which the thermal expansion is closer to the value of zirconia than La₀.₈Sr₀.₂MnO₃. This is also the composition for the

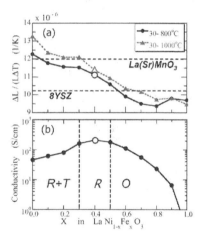

Figure 5. Composition dependence of LaNi(Fe)O₃ cathode, (a) thermal expansion coefficient and (b) conductivity at 800°C and crystal structure of LaNi(Fe)O₃. In Fig.5 (b), R, T and O indicate rhombohedral, tetragonal, orthorhombic phases, respectively.

highest electronic conductivity of $LaNi_{1-x}Fe_xO_3$. The cathodic overvoltage of $LaNi_{1-x}Fe_xO_3$ depends on the composition. We investigated the overvoltage of $LaNi_{1-x}Fe_xO_3$ cathodes at 800°C by the three terminal current interruption method, using an SASZ electrolyte supported single cell. The overvoltage exhibited its minimum value at X=0.4. This may be because higher electronic conduction is preferable for a cathodic electrochemical reaction [11]. Therefore, this composition is the best with respect to both cathodic overvoltage and electronic conductivity. This composition is also preferable in terms of the thermal expansion coefficient.

CHROMIUM POISONING DURABILITY OF LNF CATHODE

The use of a stainless steel interconnector containing Cr is an effective way to reduce cell stack production costs [15,16]. However, the conventional cathode material, La(Sr)MnO₃ (LSM), is easily poisoned by the deposition of chromium species at its triple phase boundary [17]. If the metallic interconnector surface is coated with an oxide layer, CrO_3 vaporization can be mitigated [18,19]. However, this leads to increased production costs. Therefore, the cathode material itself should be resistant to chromium poisoning. We tested the durability of LNF cathode material with respect to chromium poisoning. In Fig. 6, we plot the change in the cathode overvoltage with operating time. The measurements were performed with the three terminal current interruption method using an electrolyte supported single cell at 800°C. The IR factor was subtracted with this method. The LNF and LSM cathodes were screen-printed directly on the SASZ electrolyte sheet. An Inconel 600 metal mesh was placed near the cathodes, which were operated in air at 800°C. A conventional LSM cathode degraded rapidly with operating temperature even at low cathode overvoltage in the initial stage, but we observed no degradation of the LNF cathode at 0.2V cathode overvoltage. For comparison, results for experiments on LSM and LNF cathodes performed under almost the same conditions are also shown in Fig. 6 (a) and (b). But these cathodes were operated without the Inconel 600 mesh. The results demonstrated that an LNF cathode is much more durable than an LSM cathode as regards chromium poisoning.

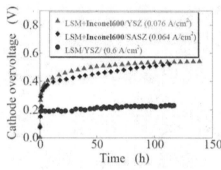 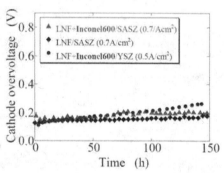

Figure 6. (a) Cathodic overvoltage IR corrected for cells using LSM cathode with/without Inconel 600 as a function of time. Triangle: LSM / YSZ / NiO-YSZ cell at 0.076A/cm², Diamond: LSM / SASZ / NiO-YSZ cell at 0.064A/cm², Circle: LSM/YSZ/NiO-YSZ cell without Inconel 600 at 0.6 A/cm².

Figure 6. (b) Cathodic overvoltage IR corrected for cells using LNF cathode as a function of time. Triangle: LNF / SASZ / NiO-YSZ cell with Inconel 600 at 0.7A/cm², Diamond: LNF/SASZ/NiO-YSZ cell without Inconel600 at 0.7A/cm², Circle: LNF / YSZ / NiO-YSZ cell with Inconel600 at 0.5A/cm².

Those cathodes were observed with EPMA after the operation. Fig. 7(a) and (b) show the Cr atom distribution at a cross section of the cathodes. The brightness corresponds to the Cr concentration. As shown in Fig. 7(b), the Cr atoms were deposited particularly at the triple phase boundary of the LSM. The chromium species are attracted to the TPB of LSM and react with LSM to from manganese-spinel phases [12-14]. But no particular chromium deposition was observed at the TPB of LNF, as shown in Fig. 7(a).Only a small amount of chromium was distributed all over the LNF layer. This showed that LNF is very stable under the cathode operating conditions including in a chromia vapor atmosphere.

Cr distribution

(a) LNF/Zr(Sc,Al)O$_2$ (b) LSM/Zr(Sc,Al)O$_2$

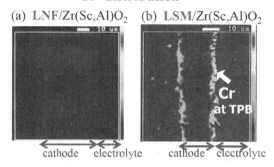

cathode   electrolyte   cathode   electrolyte

Figure 7. Chromium distribution determined by EPMA near interface of LNF cathode or LSM cathode and SASZ electrode of cells after power generation tests with Inconel 600 at 800°C for 150 h.

ACTIVE LAYER IN LNF CATHODE

If we are to apply LNF to cells of a practical size, the adhesion of the cathode layer should be sufficiently strong. The adhesion strength tends improve as the sintering temperature increases. This is because high temperature sintering promotes an interdiffusion between the cathode and the electrolyte interface. But the diffusion of La atoms from LNF into the zirconia electrolyte causes La$_2$Zr$_2$O$_7$ to form at the interface [20]. This results in a large interface resistance on the cathode side. Fortunately, the interface resistance at the LNF cathode improves greatly after operation or current application at the cathode [20,21]. This interface resistance change

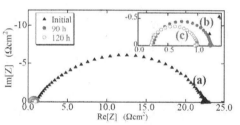

Figure 8. AC impedance plots for LNF cathode at 800°C (They were all measured at OCV). (a) Initial, after (b) 90 hours and (c)120 hours current loading (0.255 A/cm^2).

is shown in Fig. 8. The AC impedance measurements were all conducted at OCV. The measurements for (b) and (c) were done one hour after switching off the current. The diameter of the semicircle of the impedance plots represents the interface resistance. The interface resistance was very high before the current loading, but it fell to one thirtieth of its initial value. Fig. 9 plots the temporal change in the LNF cathode potential. After applying the current loading, the cathode potential decreased with the operation time. This change lasted as long as 100h. In terms of practical use, such a slow and large change in the interface resistance is unacceptable because it may result in an in plane nonhomogeneous current distribution or the nonhomogeneous performance of the cells in a stack.

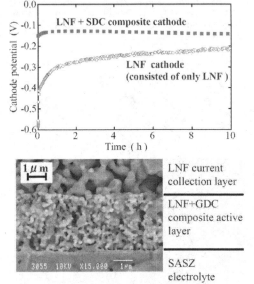

Figure 9. Cathode potential change for cells with LNF current collection layer, LNF-SDC composite active layer at 0.255A/cm^2 and 800°C. The potential for the cathode without an active layer (only LNF layer) is also plotted (open circles).

LNF current collection layer

LNF+GDC composite active layer

SASZ electrolyte

Figure 10. SEM image of a cross section near the interface between the SASZ electrolyte and the LNF cathode of a cell.

To overcome this problem, we inserted an active layer composed of rare-earth doped ceria and LNF between the zirconia electrolyte and the LNF cathode layer. Doped cerias are known to be ionic conductors that react little with most SOFC cathode materials and zirconia electrolyte [2,20]. A SEM image of a cell cross section near the active layer is shown in Fig. 10. In this layer, the ceria particle size is 0.2 μm or less in diameter, which is much smaller than that of LNF, which is about 1 μm in diameter. Small particles of ceria tend to cover the LNF particles [21]. This configuration prevents the LNF particles from direct contact with the zirconia electrolyte. Most of the LNF particles contact the electrolyte through ceria particles. This microscopic configuration in the active layer allows us to increase the sintering temperature without forming La$_2$Zr$_2$O$_7$ [20-22]. In addition to the particle size ratio of LNF and ceria, the volume ratio of the composite is also important [23]. Fig. 11 plots the total conductivity of the LNF-SDC composites with different compositions. These composites were screen printed on an Al$_2$O$_3$ sheet, and measured by the DC four terminal method in air. The conductivity of composites with 0-30wt% SDC has a temperature dependence similar to that of LNF, which is almost flat. But composites containing 50wt% or more of SDC have a steeper temperature dependence similar to that of SDC. When the SDC content is 30wt% or less, the conductivity and temperature dependence of the conductivity are quite different. At this composition, the current path changes from ionic (network consisting of ceria particles) to electronic (network consisting of LNF particles). To suppress the La$_2$Zr$_2$O$_7$ formation, we use a 50 wt% ceria composite for the active layer. A cell with this active layer exhibits no large temporal voltage change from the beginning. This is shown in Fig. 9. We investigated the sintering temperature dependence of the cathode performance and found that the optimum sintering temperature is around 1150°C, at which the adhesion of the cathode layers is practical and the cathode interface resistance becomes minimum [21]. This cathode retains its potential with 0.255mA/cm^2 operation at 800°C for more than hundred hours. The microstructure of this composite layer seems to be stable, because the sintering temperature of 1150°C is much higher than the operating temperature [21,22].

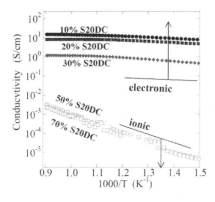

Figure 11. Temperature dependence of conductivity of screen-printed LNF-SDC composite films on Al$_2$O$_3$ substrates. The symbols represent the composition of the films: filled circles for 10wt% SDC, filled squares for 20wt% SDC, filled diamonds for 30wt% SDC, open circles for 50wt% SDC and open squares for 70wt% SDC.

ANODE SUPPORT PLANAR CELL WITH SASZ ELECTROLYTE AND LNF CATHODE

We fabricated the anode support cells using co-sintering and screen printing methods [4-6]. The photograph of as sintered cells of 60 and 120mm in diameter are shown in Fig. 12. They have an SASZ (0.89ZrO$_2$-0.1Sc$_2$O$_3$-0.01Al$_2$O$_3$) electrolyte layer, a composite SASZ-Ni anode active layer, a composite SASZ-Ni anode layer, an LNF-GDC cathode active layer and an LNF cathode layer. These SASZ-Ni composites are a mixture of 55wt% Ni and 45wt% SASZ. Fig. 13 shows a SEM image of the cross section of a cell. The green sheets for the anode layer and anode active layer were prepared by the doctor blade method. They were laminated into one sheet and then an SASZ layer was screen printed on it. The green sheets were co-sintered at 1300°C or above. The LNF(LaNi$_{0.6}$Fe$_{0.4}$O$_3$) - GDC(Ce$_{0.9}$Gd$_{0.1}$O$_2$.) composite cathode active layer was screen printed and dried. Then an LNF current collection layer was screen printed on it [21]. Finally the cells were fired at 1150°C to sinter the cathode. The anode active layer has a finer microstructure, which enables the electrolyte to be thin and pin-hole free.

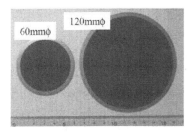

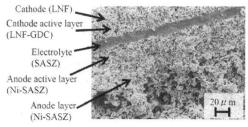

Figure 12. As sintered anode support cells 60 and 120mm in diameter.

Figure 13. SEM image of cross section of the anode support cell.

Fig. 14 shows the power generation characteristics for a 120mm diameter anode support single cell at 800°C. H_2 fuel and air were used for the anode and cathode, respectively. At 38A (0.40A/cm²), the cell voltage and power were about 0.8V and 30W. Cells 60 and 120mm in diameter yielded the same power density at any current density. We have not yet encountered any difficulty with respect to increasing the cell size. A long-term operation test was conducted for a 60mm diameter single cell with Pt current collectors at 800°C. The temporal change in the cell voltage is plotted in Fig. 15. The current density and temperature were 0.4A/cm² and 800°C, respectively. We did not observe any serious degradation over approximately 6000h that included one thermal cycle. The voltage degradation rate was 0.4% per 1000h. We plan to improve the rate to 0.25% per 1000h or less, which equates roughly to 10% over 5 years of operation.

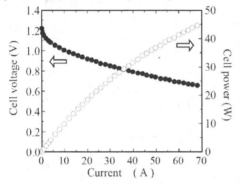

Figure 14. Single cell power generating characteristics of 120mm diameter anode support cell at 800°C.

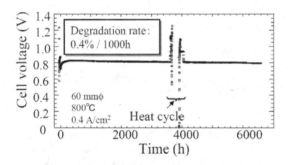

Figure 15. Durability test results for an anode support single cell (60mm in diameter) at 800°C, 0.4A/cm².

POWER GENERATION PERFORMANCE OF STACK CONSISTING OF ANODE SUPPORT PLANAR CELLS AND STAINLESS STEEL MANIFOLD

The power generation unit consists of a 120mm diameter cell and ferritic stainless steel manifold. It is illustrated in Fig. 16. The manifold contains four interconnectors (separators). Anode support planar cells are placed between the interconnectors. The fuel gas and air are supplied through the feeding path (feeding holes) and led to the cell center from where they flow to the edges of the cell. The exhaust fuel is retrieved through two exhaust paths. The exhaust air is not retrieved. This symmetrical gas flow configuration allows the fuel gas and air to reach the anode and cathode surface symmetrically [24-26]. This is advantageous as regards high fuel utilization and thus for highly efficient power generation. A contact paste and a porous metal sheet were used for the electric connections at the cathode-interconnector interface and the anode-interconnector interface, respectively [27]. A photograph of a 1.5kW class stack consisting of 40 power generation units before operation is shown in Fig. 17. We performed a durability test on this stack [28]. The temporal change in the output voltage of the 40 cell stack operated at 50A is shown in Fig. 18. The utilization ratios of dry H_2 fuel and air were 60% and 30%, respectively. There were two fuel interruption incidents at 50 and 100h caused by commercial power failures. But there was no significant degradation. This stack operated stably for 400h at 800°C.

After this test, we switched the fuel to steam reformed CH_4, and examined the power generation characteristics [28]. The steam carbon ratio of the fuel was 3 (S/C=3). The current vs. voltage (filled circles) and current vs. power (open circles) characteristics of this 40-cell stack are shown in Fig. 19. The stack furnace temperature was 800°C. This stack achieved a maximum power generation efficiency of 64% (DC, LHV) at U_{fuel}=85%. These results proved that our 40-cell stack operates stably at 50A to produce electrical power of greater than 1.5kW, and that a high electrical conversion efficiency of over 60% is attainable.

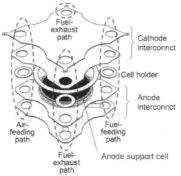

Figure 16. Construction of power generation unit. An anode supported planar cell is placed between stainless steel interconnector manifold.

Figure 17. 1.5kW class SOFC stack with LNF cathode developed by NTT. 40 node-support planar cells (120mmf) were used. Power generating efficiency was 64% at 800°C with fuel utilization rate of 85%.

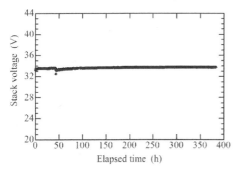

Figure 18. Temporal change of stack voltage in 40-cell stack at 50 A using dry hydrogen as fuel. The temperature of the stack furnace was 800°C.

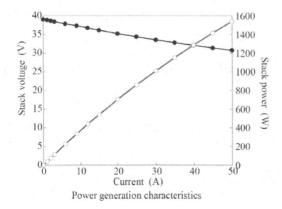

Figure 19. Current-voltage (filled circles) and power (open circles) characteristics of 40-cell stack using reformed methane (S/C=3) as fuel. The stack furnace temperature was 800°C.

Power generation characteristics

CONCLUSIONS

We succeeded in developing anode support planar cells consisting of an LNF cathode and a thin layer SASZ electrolyte and a Ni-SASZ anode. SASZ is a phase transition free zirconia electrolyte that exhibits three times the ionic conductivity of a conventional 8YSZ electrolyte at 800°C. The SASZ electrolyte layer for 120mm diameter cells is as thin as 20μm and pinhole free.

The LNF cathode was shown to be durable against chromium poisoning and to have preferential characteristics including high electronic conductivity and TEC matching with zirconia. The initial cathode performance was greatly improved by inserting the LNF-GDC composite active layer. This also made it possible for the cathode layer to be sintered at high temperature. This improved the adhesion without any deterioration in the cathode performance.

One of our planar cells was tested at $0.4A/cm^2$ and 800°C. It operated stably for about 6000h with a 0.4%/1000h degradation rate.

We adopted a stainless steel interconnectors and manifolds for our cells with high geometric symmetry, which enabled the fuel gas and air to reach the electrodes uniformly. This led to a high fuel utilization ratio and stable operation. The SOFC stack composed of 40 anode support planar cells had a power output of 1.5 kW, good durability, and a high electrical efficiency (LHV) of 64 %.

REFERENCES
[1] CSR report of NTT in 2009 (http://www.ntt.co.jp/csr_e/2009report/ecology/data.html).
[2] N. Minh, *J. Am. Ceram. Soc.*, **76**, 563-588(1993).
[3] S. C. Singhal and K. Kendall, *High-Temperature Solid Oxide Fuel Cells: Fundamentals, Design and Applications,* Elsevier Science, (2003).
[4] K. Nozawa, H. Orui, R. Chiba, and M. Arakawa, *Proceedings of the SOFC-VII*, p938 (2001).
[5] H. Orui, K. Watanabe, R. Chiba, and M. Arakawa, *J. Electrochem. Soc.*, **159(9)**, A1412-1417 (2004).
[6] H. Orui, K. Nozawa, K. Watanabe, S. Sugita, R. Chiba, T. Komatsu, H. Arai, and M. Arakawa, *J. Electrochem. Soc.*, **155**, B1110-1116 (2008).
[7] Z. Yang, K. S. Weil, D. M. Paxton, and J. W. Stevenson, *J. Electrochem. Soc.*, **150**, A1188 (2003).
[8] J. Will, A. Mitterdorfer, C. Kleinlogel, D. Perednis, and L. J. Gauckler, *Solid State Ionics*, **131**, 79 (2000).

[9] T. Ishii, *Solid State Ionics*, **78**, 333-338 (1995).

[10] R. Chiba, F. Yoshimura, and Y. Sakurai, *Solid State Ionics*, **124**, 281-288 (1999).

[11] R. Chiba, F. Yoshimura, and Y. Sakurai, *Proceedings of SOFC-VI*, **99-19**,453-462(1999).

[12] T. Komatsu, H. Arai, R. Chiba, K. Nozawa, M. Arakawa, and K. Sato, *Electrochem. and Solid-State Lett.*, **9-12(1)**, A9 (2006).

[13] T. Komatsu, H. Arai, R. Chiba, K. Nozawa, M. Arakawa, and K. Sato, *J. Electrochem. Soc.*, **154**, B379-382 (2007).

[14] T. Komatsu, R. Chiba, H. Arai, and K. Sato, *Journal of Power Sources* **176** *132–137 (2008)*.

[15] Z. Yang, K. S. Weil, D. M. Paxton, and J. W. Stevenson, *J. Electrochem. Soc.*, **150**, A1188 (2003).

[16] J. Will, A. Mitterdorfer, C. Kleinlogel, D. Perednis, and L. J. Gauckler, *Solid State Ionics*, **131**, 79 (2000).

[17] H. Tu and U. Stimming, *J. Electrochem. Soc.*, **127**, 284 (2004).

[18] Z. Yang, G. Xia, G. D. Maupin, and J. W. Stevenson, *J. Electrochem. Soc.*, **153**, A1852 (2006).

[19] M. Stanislowski, J. Froitzheim, L. Niewolak, W. J. Quadakkers, K. Hilpert, T. Markus, and L. Singheiser, *J. Power Sources*, **164**, 578 (2007).

[20] R. Chiba, T. Komatsu, Y. Tabata, H. Orui, K. Nozawa, H. Arai, and M. Arakawa, *Solid State Ionics*, **178**,1701-1709(2008).

[21] R. Chiba, H. Orui, T. Komatsu, Y. Tabata, K. Nozawa, M. Arakawa, K. Sato, and H. Arai, *J. Electrochem. Soc.*, **155**, B575-580 (2008).

[22] H. Arai, R. Chiba, T. Komatsu, H. Orui, S. Sugita, Y. Tabata, K. Nozawa, K. Watanabe, M. Arakawa, and K. Sato, *The Journal of Fuel Cell Technology*, **5**, 031204-1 - 031204-5 (2008).

[23] R. Chiba, T. Komatsu, H. Orui, H. Taguchi, K. Nozawa, and H. Arai, *Journal of the Korean Ceramic Society*, **Vol. 45**, No. 12, pp. 766~771, (2008).

[24] S. Sugita, H. Arai, Y. Yoshida, H. Orui, and M. Arakawa, *ECS Trans.* **5 (1)**, 491 (2007).

[25] M. Yokoo, Y. Tabata, Y. Yoshida, K. Hayashi, K. Nozawa, Y. Nozaki, and H. Arai, *J.Power Source*, **178/1**, 59 (2008).

[26] M. Yokoo, T. Tabata, Y. Yoshida, H. Orui, K. Hayashi, Y. Nozaki, K. Nozawa, and H. Arai, *J. Power Sources*, **184**, 84 (2008).

[27] Y. Yoshida, T. Komatsu, R. Chiba, M. Yokoo, K. Hayashi, H. Orui, and H. Arai, *ECS Transactions*, **25 (2)** 421-428 (2009).

[28] H. Arai, K. Hayashi, Y. Yoshida, M. Yokoo, Y. Mizutani, K. Ukai, K. Katsurayama, H. Orishima, M. Hirakawa, H. Uwani, and S. Kashima, **ECS Transactions, 25 (2) 125-132 (2009).**

INVESTIGATION OF THE EFFECTS OF NIO-SCSZ-LAYER INSERTION ON THE CURRENT COLLECTION AND CATALYTIC PROPERTIES OF SCSZ-BASED MICRO-TUBULAR SOFC

Toshiaki Yamaguchi[1,2] and Nigel Sammes[2]
[1]National Institute of Advanced Industrial Science and Technology
  Nagoya, Aichi, Japan
[2]Dept. of Metallurgical & Materials Engineering, Colorado School of Mines
  Golden, Colorado, USA

ABSTRACT

   This report summarizes the fabrication and evaluation of a microtubular SOFC supported by a micro anode tube, composed of Sc_2O_3-doped ZrO_2 (ScSZ) and NiO mixture. The cell was fabricated through a co-sintering of a ScSZ electrolyte, a NiO-ScSZ activation layer and a NiO-ScSZ anode support, and then a $La_{0.6}Sr_{0.4}Co_{0.2}Fe_{0.8}O_3$-x-$Ce_{0.9}Gd_{0.1}O_{1.95}$ cathode layer was deposited on the electrolyte film. Evaluation was conducted using a potentio/galvanostat and impedance analyzer under a humidified hydrogen flow. Power densities at 0.7 V were 45.1 and 67.3 mW/cm^2 at 700 and 750 C, respectively. Impedance analyses showed that with increasing operation temperature, the total resistance (ohmic + polarization) became decreased in the case of the cell with the anode activation layer, while that for the cell without the anode activation layer depended weakly upon the temperature above 700 C.

INTROSUCTION

   Recently, solid oxide fuel cell (SOFC) has received a great deal of attention, due to the high energy conversion efficiency and environmental compatibility.[1-3] The use of alternative high performance electrolytes and electrodes materials means that the SOFC operation temperature can be reduced.[4,5] In addition, miniaturization of the cell size and integration of multiple miniaturized cells is also considered to be one of the most effective approaches to increase the volumetric power density.[6] Micro tubular SOFC design exhibits many desirable characteristics, such as increasing the net electrode area per unit volume and the durability to thermal stress caused by rapid start-up operation.[7-9] Very recently, various researchers reported the excellent electrochemical performances and durability of heat-cycles for the microtubular SOFC design.[10-12] Most of the cells are supported by a porous anode tube. Therefore the anode tubular support works as a current collector and gas diffusion layer as well as support medium of SOFC. Thus, the aim of this study is to investigate the benefit of the anode current collection method on the performance of the microtubular SOFCs by testing various current collection methods. In this paper, we will report the results on feasibility study for cell preparation in order to apply investigations of the benefits of current collection methods mainly for the anode support.

15

EXPERIMENTAL

In this study, we used Sc_2O_3 doped ZrO_2 (ScSZ, Daiichi-Kigenso Kagaku Kogyo Co. Ltd., Japan), NiO (NiO-F grade, Fuel Cell Materials, USA) as raw materials for an anode and an electrolyte. The samples were prepared via a co-sintering of a ScSZ electrolyte layer and 50wt% NiO-50wt% ScSZ anode microtubular support, and then an application of a commercial 50wt% $(La_{0.6}Sr_{0.4})(Co_{0.2}Fe_{0.8})O_{3-x}$ (LSCF)-50wt% $(Ce_{0.9}Gd_{0.1})O_{1.95}$ (GDC) cathode ink (Fuel Cell Materials, USA) on the electrolyte film. Initially, a NiO-ScSZ tubular support was extruded using a mixture of NiO and ScSZ powders, methylcellulose binder (Type A E461 Modified Cellulose, DOW, USA) and water. The extruded tubular support has a diameter approximately 2.4 mm with a fuel gas channel of approximately 1.6 mm in diameter. The green anode tube was dip-coated with a NiO-ScSZ slurry to activate the anode reaction and then calcined at 1000 C for 2 h in air. The NiO-ScSZ slurry was prepared by ball-milling the solid contents (60wt% NiO/ 40wt% ScSZ powder mixture, surface modifier (M1201, FERRO, USA), butyral binder (B98, BUTVAR, USA) and plasticizer (triethylene glycol di-2-ethylhexanoate, BUTVAR, USA)) and an ethanol/toluene mixed solvent for 72 h. For comparison, an anode tube without the NiO-ScSZ activation layer was also calcined at 1000 C for 2 h in air. The calcined microanode tubes were coated by a ScSZ slurry and then co-sintered at 1300 C for 2 h in air. The ScSZ slurry was also prepared by ball-milling the solid contents (ScSZ powder, surface modifier, butyral binder and plasticizer) and the ethanol/toluene mixed solvent.

Next, the ScSZ-coated tubes were applied with the commercial LSCF-GDC cathode ink, and then the tubes were heated at 1000 C for 2 h in air. The completed tubular cell has a diameter of approximately 1.6 mm and a cathode length of approximately 10 mm, that is, the effective electrode area is approximately 0.5 cm^2. The microtubular SOFCs with and without the NiO-ScSZ activation layer were denoted as *cells A* and *B*.

The performances of the cell were measured from 650 to 750 C using a Chroma 6310 series for current-voltage characterization, and a Gamry 600 for impedance analyses under humidified H_2 (3% H_2O) gas flow at a flow rate of 25 cm^3/min and open air. Ag wire was used for current collection from the anode and cathode sides, and the wires on both sides were fixed using Ag paste (903-B; ESL ElectroScience). Current collection from the anode was conducted by winding a Ag wire around the edge of the anode support. The impedance measurements were conducted in a frequency range from 0.1 Hz to 100 kHz with a signal amplitude of 5 mV under open-circuit condition. The detailed experimental apparatus of the single cell measurement was described elsewhere.[13]

RESULTS AND DISCUSSIONS

Figure 1 shows the voltages and power densities of the *cells A* and *B* as a function of current density at 700 and 750 C. The open circuit voltages (OCVs) of both of cells are above 1.07 V. A commercial ceramic bond is used as the gas sealant for the single-cell measurements. In view of gas leakage through the ceramic bond layer with partial gas

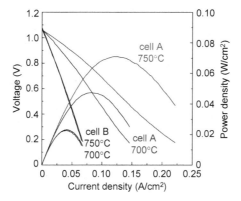

Fig.1 Voltages and power densities of cells A and B as a function of current density

tightness, it is considered that the ScSZ electrolyte was enough densified. The I-V curves in the *cell B* dropped rapidly at low current densities. The power densities at 0.7 V were 45.1 and 67.3 mW/cm^2 for the *cell A*, and 21.5 and 21.6 mW/cm^2 for the *cell B* at 700 and 750 C, respectively, so the *cell A* can generate a power density more than 3 times higher than the *cell B*. And, the maximum power densities were 47.4 and 71.0 and 22.5 mW/cm^2 for the *cell A*, and 22.5 and 23.3 mW/cm^2 for the *cell B* at 700 and 750 C, respectively, so the maximum power density of the *cell A* is also more than 3 times that of the *cell B*.

Figure 2 summarizes the impedance spectra of the *cells A* and *B* from 650 to 750 C. Total resistance (ohmic + polarization) for the *cell A* is approximately 3 times larger

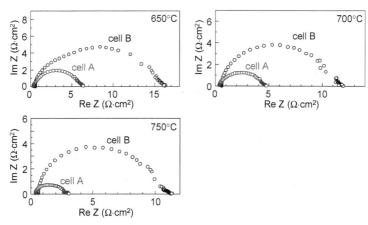

Fig.2 Impedance spectra of cells A and B at various temperatures

than that for the *cell B*, which is correspond with the result in Fig. 1. In the case of the *cell A*, the total resistances decreased with increasing the temperature. On the other hand, the total resistance for the *cell B* decreased from 650 C to 700 C; however, the values at 700 and 750 C were almost the same.

Figure 3 shows the temperature dependence of the area specific resistances for the ohmic and polarization. The ohmic resistances for the *cells A* and *B* had a flat tendency to the temperature, and the values of the *cells A* and *B* were approximately 0.35 and 0.60 cm², respectively. Compared to the temperature dependence of the ohmic resistance of usual electrolyte materials, the ohmic resistances in this study was weakly depended on the temperature, which indicates that the in-plane ohmic resistance for the tubular anode support is dominant in the total ohmic resistance. And, the lower ohmic resistance for the *cell A* indicates that the activation layer improved the in-plane conduction of the current at the anode tube. On the other hand, the electrode polarizations for the *cells A* and *B* followed Arrhenius plot as can be seen in Fig. 3, which shows that the semi-circle in lower frequency region mainly attributed electrode activation energy. The weaker temperature dependence of the electrode polarization for the cell B causes the almost the same performances of the I-V and impedance curves, as can be seen in Figs. 1 and 2. However, the *cell A* still has a large ohmic and polarization resistances, currently we are trying to improve the cell performance by optimizing the anode microstructure and composition in order to decrease ohmic resistance and to increase anode activity (amount of three phase boundary).[9, 14]

Figure 4 shows the overpotentials of the *cells A* and *B* as a function of current density at 700 to 750 C. The overpotential for the *cell B* increased steeply even in the low current density region, and the two lines at 700 and 750 C were approximately consistent, as described above. However, the overpotentials for the *cell A* showed more gradual sloping lines. The values under 50 mA/cm² operation were 0.264 V and 0.163 V for the *cell A*, and 0.606 V and 0.603 V for the

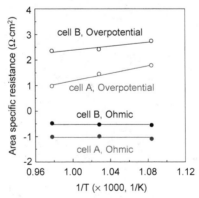

Fig.3 Area specific resistances of cells A and B as a function of reciprocal of temperature

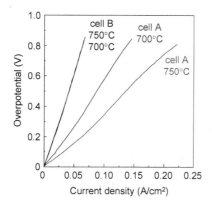

Fig.4 Overpotentials of cells A and B as a function of current density

cell B at 700 and 750 C, respectively. In view of the impedance analyses (Figs. 2 and 3), these overpotentials were mainly due to a lack of the electrode activity at the anode side.

CONCLUSION

The performance of the ScSZ-based micro-tubular SOFC was investigated and showed power densities at 0.7 V of 45.1 and 67.3 mW/cm^2 at 700 and 750 C, respectively. From the impedance analyses, the activation resistance followed Arrhenius relationships; however while the ohmic resistance depended weakly upon the temperature.

ACKNOWLEDGEMENT

This study was supported by the Japan Society for the Promotion of Science (JSPS); Excellent Young Researcher Overseas Visit Program. I am grateful to Daiichi-Kigenso Kagaku Kogyo Co. Ltd. for supplying the ScSZ powder.

REFERENCES
[1]M. Dokiya, SOFC System and Technology, *Solid State Ionics*, **152-153**, 383-92 (2002).
[2]T. Hibino, A. Hashimoto, M. Yano, M. Suzuki, S. Yoshida and M. Sano, High Performance Anodes for SOFCs Operating in Methane-Air Mixture at Reduced Temperatures, *J. Electrochem. Soc.*, **149**, A133-36 (2002).
[3]T. Ishihara, J. Tabuchi, S. Ishikawa, J. Yan, M. Enoki and H. Matsumoto, Recent Progress in LaGaO$_3$ Based Solid Electrolyte for Intermediate Temperature SOFCs, *Solid State Ionics*, **177**, 1949-53 (2006).
[4]B. C. H. Steele and A. Heinzel, Materials for Fuel Cell Technology, *Nature*, **414**, 345-52 (2001).

[5]J. W. Fergus, Electrolytes for Solid Oxide Fuel Cells, *J. Power Sources*, **162**, 30-40 (2006).

[6]N. M. Sammes, Y. Du and R. Bove, Design and Fabrication of a 100W Anode Supported Micro Tubular SOFC Stack, *J. Power Sources*, **145**, 428-34 (2005).

[7]Y. Funahashi, T. Shimamori, T. Suzuki, Y. Fujishiro and M. Awano, Optimization of Configuration for Cube Shaped SOFC Bundles, *ECS Transactions*, **7**, 643-49 (2007).

[8]T. Yamaguchi, S. Shimizu, T. Suzuki, Y. Fujishiro and M. Awano, Fabrication and Characterization of High Performance Cathode Supported Small Scale SOFC for Intermediate Temperature Operation, *Electrochem. Commun.*, **10**, 1381-83 (2008).

[9]T. Suzuki, Z. Hasan, Y. Funahashi, T. Yamaguchi, Y. Fujishiro and M. Awano, Impact of Anode Microstructure on Solid Oxide Fuel Cells, *Science*, **325**, 852-55 (2009).

[10]I. P. Kilbride, Preparation and Properties of Small Diameter Tubular Solid Oxide Fuel Cells for Rapid Start Up, *J. Power Sources*, **61**, 167-71 (1996).

[11]K. Kendall and M. Palin, A Small Solid Oxide Fuel Cell Demonstrator for Microelectronic Application, *J. Power Sources*, **71**, 268-70 (1998).

[12]K. Yashiro, N. Yamada, T. Kawada, J. Hong, A. Kaimai, Y. Nigara and J. Mizusaki, Demonstration and Stack Concept of Quick Startup/Shutdown SOFC, *Electrochemistry*, **70**, 958-60 (2002).

[13]T. Yamaguchi, K. Galloway, J. Yoon and N. Sammes, Electrochemical Characterizations of Microtubular Solid Oxide Fuel Cells under a Long Term Testing at Intermediate Temperature Operation, *J. Power Sources*, **196**, 2627-30 (2011).

[14]T. Yamaguchi, S. Shimizu, T. Suzuki, Y. Fujishiro and M. Awano, Effects of Anode Microstructure on the Performances of Cathode-Supported Micro SOFCs, *Electrochem. Solid-State Lett.*, **12**, B151-53 (2009)

Electrolytes

EFFECT OF DOPANTS ON CeO$_2$ BASED SOLID STATE ELECTROLYTES FOR
INTERMEDIATE TEMPERATURE ELECTROCHEMICAL DEVICES

E. Yu. Pikalova and A. K. Demin
Institute of High Temperature Electrochemistry, Ural's Brunch of Russian Academy of Science
Ekaterinburg, Russia

V. G. Bamburov
Institute of Solid State Chemistry, Ural's Brunch of Russian Academy of Science Ekaterinburg,
Russia.

V.I. Maragou and P.E. Tsiakaras
Department Mechanical Engineering, School of Engineering, University of Thessaly
Volos, Greece.

ABSTRACT
The present work aims at the investigation of the influence of different dopants' ionic radii and their
concentration on both the lattice parameters and the density of Ce$_{1-x}$Ln$_x$O$_{2-\delta}$ (x = 0-0.20; Ln = La, Nd, Sm, Eu,
Gd, Dy, Ho, Er, Yb) solid solutions, multi-component Ce$_{1-x}$Ln$_{x/2}$Ln'$_{x/2}$O$_{2-\delta}$ (x = 0-0.20; Ln = Sm, La, Gd and
Ln' = Dy, Nd, Y), Ce$_{0.8}$(Sm$_{1-x}$M$_x$)$_{0.2}$O$_{2-\delta}$ (M= Ca, Sr; x=0.0-1.0) and Ce$_{0.8}$(Sm$_{1-x-y}$Ba$_y$M$_x$)$_{0.2}$O$_{2-\delta}$ (M=Ca, Sr;
x=0, 0.15, 0.20; y=0.05, 0.1) systems. The electrical conductivity measurements of the samples were
carried out by using the four-point dc technique in different atmospheres. Temperature and oxygen
partial pressure were varied automatically by means of microprocessor system ZIRKONIA-318 in the
range of 623 – 1173 K and 0.1-10^{-23} atm, respectively. To describe the electrolytic properties of solid
electrolytes the notation of the EDB (electorytic domain boundary) - the critical oxygen partial
pressure ($P_{O_2}^*$) at which the values of the electronic and ionic components of conductivity are equal- is
often used. The values of the critical partial pressure of oxygen for different compositions at 1023 K
were calculated from dependences of total conductivity vs. oxygen partial pressure in proposal that
ionic conductivity depends on dopand's concentration, but isn't influenced by P_{O_2}.

INTRODUCTION
        Doped cerium dioxide materials, Ce$_{1-x}$M$_x$O$_{2-\delta}$ (M: rare-earth or alkaline-earth cations), have been
attracting the increasing interest as electrolytes in intermediate temperature solid oxide fuel cells, due to their higher
ionic conductivity with respect to stabilized zirconia and their lower cost in comparison with lanthanum gallate-
based materials [1]. The influence of dopant on the electrical properties of solid electrolytes based on CeO$_2$ was
examined in several studies both experimental [2-5] and theoretical [6-8], and reviewed since 1996 [9-11]. Some
investigations dealing with the multi-component electrolytes on the base of CeO$_2$ [12-16] and brief reviews [17]
should be noted here. However, despite the abundance of information, we did not find a work, in which important
properties of solid electrolytes based on CeO$_2$, doped with rare-earth or alkaline-earth elements, such as the total
conductivity in air atmosphere in a wide range of temperatures and dopant concentrations and the electrolytic
domain boundary is presented. Since these properties of the electrolyte depend on such factors as the method of
synthesis and sintering regime, the comparison of the characteristics listed in the works of different research groups
are often ineffective. The present work aims at the investigation of the influence of different dopants' concentration
on the total electrical conductivity of Ce$_{1-x}$Ln$_x$O$_{2-\delta}$ (x = 0.15-0.20; Ln = La, Nd, Sm, Eu, Gd, Dy, Ho, Er, Yb),
multi-component Ce$_{1-x}$Ln$_{x/2}$Ln'$_{x/2}$O$_{2-\delta}$ (x = 0-0.20; Ln = Sm, La, Gd and Ln' = Dy, Nd, Y) and Ce$_{(1-x-y)}$Sm$_x$M$_y$O$_{2-\delta}$ (M= Ca, Sr, Ba) compositions in a wide range of temperatures in the air atmosphere, as well as in
a wide range of oxygen partial pressures at 1023 K.

23

EXPERIMENTAL

The solid state compositions were synthesized via a solid state reaction from oxides of high purity [18-19]. The powders were weighed according to the stoichiomertry and ball milled in ethyl alcohol medium for 1 h in a plastic bottle using zirconia balls in the planetary mill FRITSCH pulverisette-6. The sintering involved several stages. First, the mixtures were held at 1173 K then the reaction masses were cooled, ground for 2 h and calcined again at 1323 K for 10 h. The substances thus obtained were ground for 2 h.

The specific surface area, tested by particle size analyzer SA-CP2 Shimadzu, was approximately 1.92 m^2/g. The samples were prepared in the form of plates by rolling with the addition of 5% solution of natural rubber in the mixture of acetone and gasoline (60/40). The samples of size about 20 5 1 mm were cut from the plates with subsequent calcinations for elimination of organic binder at 873 K for 5 h with a heating/cooling rate of 0.5 deg/min. Then the samples were sintered at 1823-1923 K for 3 h in air with a heating/cooling rate of 2 deg/min. The density of the sintered samples was measured by employing both the Archimedes principle with water and the calculation from the weights and the dimensions of the specimens [18]. It was found that both methods of obtaining the density provided almost the same value. The samples with relative densities of no less then 92-95% were chosen for the electrical conductivity measurements.

XRD was performed on the milled powders of specimens after sintering by using a DMAX-2500, Rigaku Co. Ltd., Japan diffractometer with Ni-filtered CuKα radiation in the range of $10° \le 2$ ° $\le 120°$. Phase identification performed using JCPDS card file and software package Peak Find v1.03. The lattice parameters of the samples were refined using the Rietveld profile analysis method with the Fullprof program (sp. gr. $Fm3m$), as described in [19].

The electrical conductivity measurements of the samples were carried out by the four-point dc technique in different atmospheres. Temperature and oxygen partial pressure were varied automatically by means of microprocessor system ZIRKONIA-318 in the range of 673-1173 K and $0.1-10^{-23}$ atm, respectively.

RESULTS AND DISCUSSION
Samples' characterization

The XRD patterns at room temperature of the $Ce_{1-x}Ln_xO_{2-\delta}$ (x = 0.20) [18] and $Ce_{1-x}Ln_{x/2}Ln'_{x/2}O_{2-\delta}$ (x = 0.20; Ln = Sm, La, Gd and Ln' = Dy, Nd, Y) samples are presented in Figs 1a [18] & 1b. All the investigated compositions are single phase with a cubic fluorite structure like pure CeO_2. The concentration dependences of the lattice's parameters of the solid solutions are shown in Figs 2a [18] & 2b. The error limit was estimated to be $\pm$ 0.0002 nm. For the description of the lattice's parameters, a, of $Ce_{1-x}Ln_xO_{2-\delta}$ solid solutions, the following linear empirical equation is suggested

$$a = 0.5414 - \Sigma(0.0240\Delta r_k - 0.00015\Delta z_k)m_k \qquad (1)$$

where Δr_k and Δz_k - the difference between radii and valences of k-dopant and Ce^{4+}, m_k - molar content of k-dopand. The experimental values of the lattice's parameters for all $Ce_{1-x}Ln_{x/2}Ln'_{x/2}O_{2-\delta}$ compositions are in a good agreement with (1), except for $Ce_{1-x}La_{x/2}Dy_{x/2}O_{2-\delta}$. The solid solutions with La/Dy and Sm/Nd as the dopants have the same values of average dopant radii, calculated in agreement with the values of effective ionic radii of single dopants [20]; however, the lattice parameter of La/Dy is considerably lower than that of Sm/Nd.

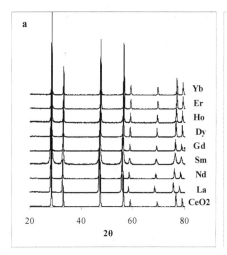

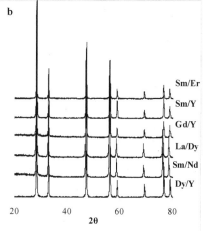

Figure 1. XRD patterns of the Ce$_{0.8}$Ln$_{0.2}$O$_{2-\delta}$ [18] (a) and the Ce$_{0.8}$Ln$_{0.1}$Ln'$_{0.1}$O$_{2-\delta}$ (b) solid solutions

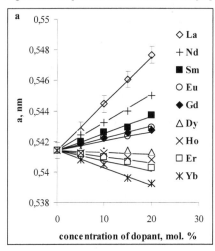

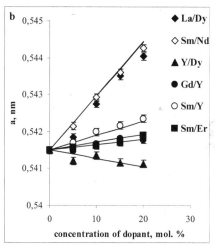

Figure 2. Concentration dependences of lattice's parameters of the Ce$_{0.8}$Ln$_{0.2}$O$_{2-\delta}$ [18] (a) and the Ce$_{0.8}$Ln$_{0.1}$Ln'$_{0.1}$O$_{2-\delta}$ (b).

As far as it concerns the compositions Ce$_{0.8}$(Sm$_{1-x}$M$_x$)$_{0.2}$O$_{2-\delta}$ (M= Ca, Sr; x=0.0-1.0), the ones with Ca^{2+} are single phase, while due to the fact that the solubility limit of Sr^{2+} is ~8 mol. %, so at x 0.4 the second phase SrCeO$_3$ was identified. All the XRD-patterns of the Ce$_{0.8}$(Sm$_{1-x-y}$Ba$_y$M$_x$)$_{0.2}$O$_{2-\delta}$

(M=Ca, Sr; x=0, 0.15, 0.20; y=0.05, 0.1) samples showed a pure cubic structure without any second phases. However, the microstructure of these samples obtained by scanning electron microscopy, demonstrated the appearance of second phases in the triple points [21]. In Figs 3a & 3b the concentration dependences of the lattice's parameters of Ce$_{0.8}$(Sm$_{1-x}$M$_x$)$_{0.2}$O$_{2-\delta}$ (M= Ca, Sr; x=0.0-1.0) are depicted. The theoretical values of the parameters in the frame of hard sphere model were calculated with the assumption of the different values of oxygen vacancy radii for two- and three-valence dopants [22-23].

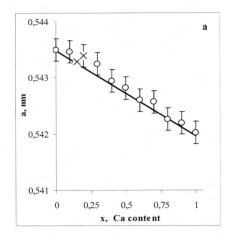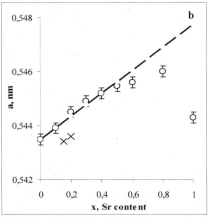

Figure 3. Concentration dependences of cell's parameters of the Ce$_{0.8}$(Sm$_{1-x}$M$_x$)$_{0.2}$O$_{2-\delta}$ (○) and the Ce$_{0.8}$(Sm$_{1-x-0.05}$Ba$_{0.05}$M$_x$)$_{0.2}$O$_{2-\delta}$ (✗solid solutions, where M = Ca (a); Sr(b).

Total electrical conductivity in the air

The total electrical conductivity is the most important property of solid state electrolytes for electrochemical applications. From this point of view, it would be interesting to study the influence of the nature of dopant and doping level on the properties of these materials, prepared in the same conditions, and to find out whether co-doping would give reasonable effect here. Fig. 4 presents the total conductivity in the air of Ce$_{0.8}$Ln$_{0.2}$O$_{2-\delta}$ solid solutions in high (873-1173 K) and low (673-873 K) temperature ranges. As the transition point between the temperature's ranges a critical temperature T$_c$ of 873 K was taken. It is known that below T$_c$ (for Ce$_{0.8}$Sm$_{0.2}$O$_{2-\delta}$ the value of T$_c$ according to calorimetric measurements [24] is 880 K, while for Ce$_{0.9}$Gd$_{0.1}$O$_{1.95}$ - 856 ± 45 K [25]) the formation of the dopant cation - oxygen vacancies complexes plays a significant role. The dopant cations may not only act just as traps for isolated vacancies, but also as nuclei for the formation of ordered-vacancy clusters. In accordance with this assumption the Arrenius plot for ceria based systems would gives two straight lines, which intersects at T$_c$ with activation energies E$_1$=ΔH and E$_1$=ΔH$_m$ +ΔH$_t$, where ΔH$_m$ – the migration energy of oxygen vacancies, ΔH$_t$ - the trapping energy between vacancies and cations [18].

The introduction of the dopant cation (regardless of its radius) significantly increases the total electrical conductivity of CeO$_2$. The highest value of the electrical conductivity in the air was found for Ce$_{0.8}$Nd$_{0.2}$O$_{2-\delta}$ (14.8 ×10^{-2} and 13.3 ×10^{-3} Sm/cm, at 1173 K and 873 K, respectively). The dependence of the electrical conductivity on the effective radius of the dopant passes through a maximum at r$_d$ = 0.1109 nm.

According to the classical theory [26], the minimal distortion in the lattice and, consequently, the maximum electrical conductivity corresponds to the cation dopant with a radius of 0.1024 nm and further increase of dopant's radius will embarrass the oxygen anion movement of the diffusion paths overlapping by neighboring cations. Indeed, we obtained the minimal changes in the lattice's parameters in CeO$_2$ doped with Dy^{3+} with the effective radius of 0.1027 nm, but the electrical conductivity of samples with this dopant was rather low. Thus, to choose a suitable dopant it is necessary to take into account such factors as the size of the cation and the formation of local structures (cation-vacancies). At low temperatures the values of the electrical conductivity of the compositions with Nd^{+3} and Sm^{+3} are about the same (Fig. 4, b). It was shown on the base of DFT-calculations [27] that these dopants have equivalent values of binding energy in CeO$_2$ lattice and the bulk conductivity of solid solutions based on them are almost equal. Therefore, the values of the electrical conductivity of these solid solutions obtained by different authors are determined, apparently, by the conductivity of grain boundaries.

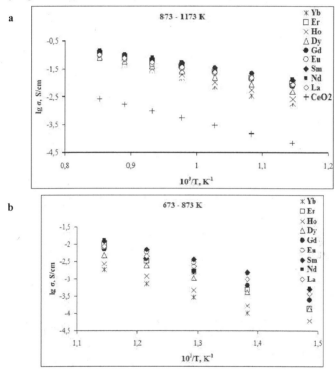

Figure 4. Total conductivity in air of Ce$_{0.8}$Ln$_{0.2}$O$_{2-\delta}$ at 873-1173 K (a) and 673-873 K (b).

Among of Ce$_{1-x}$Ln$_{x/2}$Ln'$_{x/2}$O$_{2-\delta}$ the highest values of conductivity for Ce$_{0.8}$La$_{0.1}$Dy$_{0.1}$O$_{2-\delta}$ and Ce$_{0.8}$Nd$_{0.1}$Sm$_{0.1}$O$_{2-\delta}$ were observed (11.1 $\times 10^{-2}$ and 10.0 $\times 10^{-2}$ Sm /cm at 1173 K) (Fig. 5), but an effect of increasing of conductivity under co-doping of CeO$_2$ by two rare earth elements was not confirmed. As a rule the values of conductivity of the three-component systems are equal to the average value of conductivities of the relevant two-component systems. In fig. 6 the concentration dependences of total conductivity at 1173 K for Ce$_{1-x}$Ln$_x$O$_{2-\delta}$ solid solutions and multi-component Ce$_{1-x}$Ln$_{x/2}$Ln'$_{x/2}$O$_{2-\delta}$ systems are presented. Throughout the concentration range for two-component systems the maximum of conductivity was found with Nd^{3+} as a dopant (Fig. 6a). For all the compositions studied, regardless of the radius of the dopant, the electrical conductivity increases in the range of concentrations from 10 to 20 % mol. On the contrary with the results of some authors [13, 14, 28] we did not observe an increase of the electrical conductivity of Ce$_{1-x}$Ln$_{x/2}$Ln'$_{x/2}$O$_{2-\delta}$ in comparison with the two-component systems throughout the whole range of the investigated concentrations (Fig. 6b). These results are similar to those obtained in previous work [12]. It was found that the co-doping with two or more rare earth elements affects positively mainly the conductivity in the concentration range of 10-12 mol. % and relatively low temperatures.

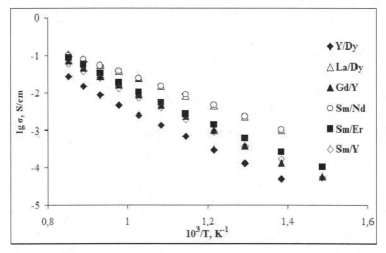

Figure 5. Total conductivity in air of Ce$_{0.8}$Ln$_{0.1}$Ln'$_{0.1}$O$_{2-\delta}$ at 673-1173 K.

Fig. 7 shows the temperature dependences of the total conductivity in air of Ce$_{0.8}$(Sm$_{1-x}$M$_x$)$_{0.2}$O$_{2-\delta}$ (M= Ca, Sr; x=0.0-1.0). The addition of the proper amount of alkaline earth elements Ca^{2+} and Sr^{2+} (2 – 4 mol. %) increases the conductivity of samples significantly. The same results were found for Ce$_{0.8}$(Sm$_{1-x-y}$Ba$_y$M$_x$)$_{0.2}$O$_{2-\delta}$ (M=Ca, Sr; x=0, 0.15, 0.20; y=0.05, 0.1) system (Table 1). It was confirmed by the technique of the impedance spectroscopy (these results are not presented here) that CaO, SrO and BaO addition increases the grain boundary conductivity more than ten times of magnitude. The results are in a good agreement with the experimental results reported in literature [29-32]. It should be noted that the conductivity of the samples with Sr^{2+} is higher due to lower binding energy of Sr-O (33.4 eV) compared to Ca-O (35.44 eV) [21]. The binding energy of Ba-O is minimal among the alkaline earth elements (31.44 eV), but due to the fact that the effective radius of Ba^{2+} is 0.142 nm, and its solubility in the fluorite lattice is less than 2 mol. % [1], the number of single-phase solid solutions on the base of BaO is limited. Nevertheless, it was found that the samples with Ba^{2+} have the lowest activation energy, for the total conductivity (Table 1).

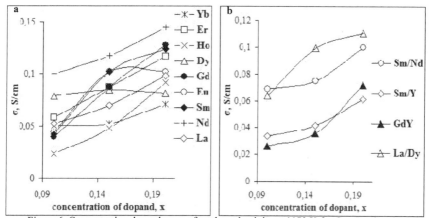

Figure 6. Concentration dependences of total conductivity at 1173 K for Ce$_{1-x}$Ln$_x$O$_{2-\delta}$ (a) and Ce$_{1-x}$Ln$_{x/2}$Ln'$_{x/2}$O$_{2-\delta}$ (b)

Table I. Electrical properties of Ce$_{0.8}$Sm$_{0.2}$O$_{2-\delta}$, co-doped with CaO, SrO and BaO.

Samples	$\sigma \cdot 10^2$, Sm/cm (1173 K)	$\sigma \cdot 10^3$, Sm/cm (873 K)	Ea, eV (873-1173 K)	Ea, eV (623-873 K)
Ce$_{0.8}$Sm$_{0.2}$O$_{2-\delta}$	11.62	3.9	1.10	0.96
Ce$_{0.8}$(Sm$_{0.8}$Ca$_{0.2}$)$_{0.2}$O$_{2-\delta}$	154.9	13.7	0.80	0.87
Ce$_{0.8}$(Sm$_{0.6}$Ca$_{0.4}$)$_{0.2}$O$_{2-\delta}$	152.7	11.8	0.84	0.92
Ce$_{0.8}$(Sm$_{0.5}$Ca$_{0.5}$)$_{0.2}$O$_{2-\delta}$	127.4	9.6	0.85	0.82
Ce$_{0.8}$(Sm$_{0.5}$Ca$_{0.5}$)$_{0.2}$O$_{2-\delta}$ [30]		48.0	0.65	
Ce$_{0.8}$(Sm$_{0.5}$Ca$_{0.5}$)$_{0.2}$O$_{2-\delta}$ [32]		7.31	0.80	
Ce$_{0.8}$Ca$_{0.2}$O$_{2-\delta}$	77.9	5.1	0.89	0.85
Ce$_{0.8}$(Sm$_{0.9}$Sr$_{0.1}$)$_{0.2}$O$_{2-\delta}$	162.1	15.2	0.78	0.82
Ce$_{0.8}$(Sm$_{0.8}$Sr$_{0.2}$)$_{0.2}$O$_{2-\delta}$	162.2	14.8	0.79	0.87
Ce$_{0.8}$(Sm$_{0.7}$Sr$_{0.3}$)$_{0.2}$O$_{2-\delta}$	150.8	12.4	0.82	0.84
Ce$_{0.8}$(Sm$_{0.9}$Ba$_{0.1}$)$_{0.2}$O$_{2-\delta}$	122.8	12.3	0.76	0.79
Ce$_{0.8}$(Sm$_{0.8}$Sr$_{0.15}$Ba$_{0.05}$)$_{0.2}$O$_{2-\delta}$	148.2	12.8	0.81	0.88
Ce$_{0.8}$(Sm$_{0.75}$Sr$_{0.2}$Ba$_{0.05}$)$_{0.2}$O$_{2-\delta}$	161.1	16.1	0.76	0.81
Ce$_{0.8}$(Sm$_{0.8}$Ca$_{0.15}$Ba$_{0.05}$)$_{0.2}$O$_{2-\delta}$	113.4	9.7	0.80	0.88
Ce$_{0.8}$(Sm$_{0.75}$Ca$_{0.2}$Ba$_{0.05}$)$_{0.2}$O$_{2-\delta}$	142.6	11.9	0.82	0.71

The dependence of the total electrical conductivity on oxygen partial pressure
    It is well known that the rate of electrochemical reactions in solid oxide fuel cells is determined by the driving forces of the processes, which, in turn, are determined by the difference of partial pressure of oxygen in the anode and cathode channels. The greater this difference, the higher the power output of electrochemical cell is. Usually, with the cathode channel oxygen partial pressure is chosen in the range 0.21-1 atm. Problems arise in reducing conditions (anode channel), since at the partial pressure of oxygen below 10^{-15} atm cerium systems are characterized by the appearance of electronic conductivity due to the partial reduction of Ce^{4+} →

Ce^{3+}. The electronic conductivity of solid solutions based on CeO$_2$ is determined by the doping level and temperature [10]. The dependences of the total electrical conductivity of Ce$_{0.8}$Ln$_{0.2}$O$_{2-\delta}$ and Ce$_{0.8}$Ln$_{0.1}$Ln'$_{0.1}$O$_{2-\delta}$ on partial pressure at T = 1073 are shown in in Figs 8a & 8b, respectively.

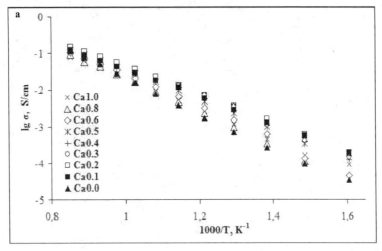

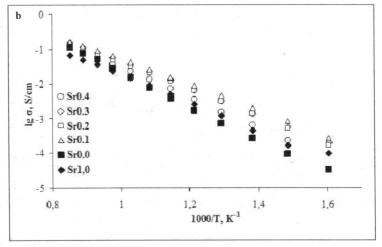

Figure 7. Total conductivity in air of Ce$_{0.8}$(Sm$_{1-x}$M$_x$)$_{0.2}$O$_{2-\delta}$ M= Ca (a), Sr (b)

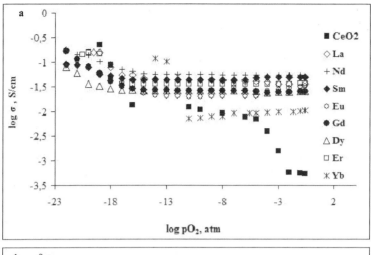

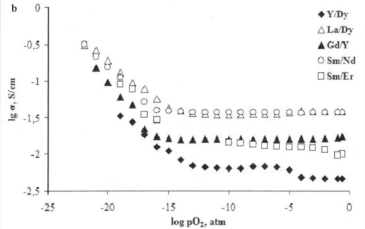

Figure 8. The effect partial pressure on total electrical conductivity of Ce$_{0.8}$Ln$_{0.2}$O$_{2-\delta}$ (a) and Ce$_{0.8}$Ln$_{0.1}$Ln'$_{0.1}$O$_{2-\delta}$ (b), at T = 1023 K

To describe the electrolytic properties of the solids the critical oxygen partial pressure $p_{O_2}^*$ at which the values of the electronic and ionic component of conductivity are equal was used. The lowest value of $p_{O_2}^*$ at 1073 K corresponds to the composition Ce$_{0.8}$Sm$_{0.2}$O$_{2-\delta}$ 2.76·10^{-21} atm, which is in good agreement with results reported in literature [2].

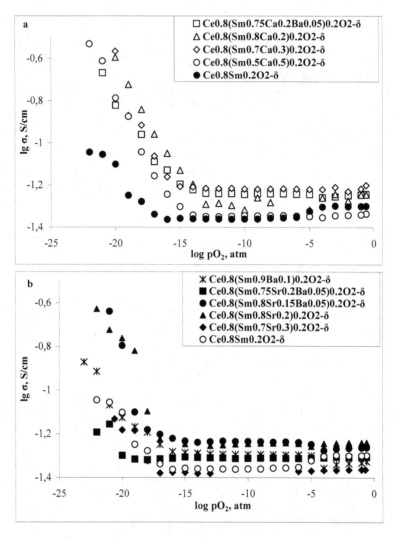

Figure 9. The effect of partial pressure on the total electrical conductivity of Ce$_{0.8}$(Sm$_{1-x}$M$_x$)$_{0.2}$O$_{2-\delta}$ and Ce$_{0.8}$(Sm$_{1-x-y}$Ba$_y$M$_x$)$_{0.2}$O$_{2-\delta}$ M= Ca (a), Sr (b), at T = 1023 K

Among the three-component systems the lowest values of $p_{O_2}^*$ equal 1.16×10^{-18} atm was found for Ce$_{0.8}$Nd$_{0.1}$Sm$_{0.1}$O$_{2-\delta}$. For compositions Ce$_{0.8}$(Sm$_{1-x}$M$_x$)$_{0.2}$O$_{2-\delta}$ the value of $p_{O_2}^*$ increases with CaO content (with x

= 0; 0.2 and 0.5 are $2.76\cdot10^{-21}$; $5.05\cdot10^{-18}$; $1.31\cdot10^{-18}$ atm, respectively), while SrO addition decreases it (with x = 0.2 and 0.3 are $3.33\cdot10^{-19}$ and $9.88\cdot10^{-23}$ atm) (Fig. 9). The addition of BaO also leads to expansion of electrolytic area of co-doped systems. The lowest value $3.30\cdot10^{-23}$ atm was found for $Ce_{0.8}(Sm_{0.75}Sr_{0.2}Ba_{0.05})_{0.2}O_{2-\delta}$. In some investigations of Mori group [33] it was shown that co-doping with an alkali and an alkaline earth elements not only leads to the increase of electrical conductivity of solid electrolytes based on CeO$_2$, but also expands their electrolytic areas.

CONCLUSIONS

The effect of cation radius of the dopant and its concentration on the physicochemical properties of solid solutions $Ce_{1-x}Ln_xO_{2-\delta}$ (x = 0 - 0.20; Ln = La, Nd, Sm, Eu, Gd, Dy, Ho, Er, Yb) was investigated. It was found that in the whole concentration range compositions with Nd^{3+} possess the highest values of conductivity at high temperatures (873-1173 K), while at low temperatures (below 873 K) – the ones with Sm^{3+} do. The investigation of multi-component systems $Ce_{1-x}Ln_{x/2}Ln'_{x/2}O_{2-\delta}$ (x = 0.20; Ln = Sm, La, Gd and Ln' = Dy, Nd, Y), $Ce_{0.8}(Sm_{1-x}M_x)_{0.2}O_{2-\delta}$ and $Ce_{0.8}(Sm_{1-x-y}Ba_yM_x)_{0.2}O_{2-\delta}$ (M=Cs, Sr) showed that the increase of the conductivity is observed only with the addition of alkaline earth elements. The highest value of the total conductivity $16.62\cdot10^{-2}$ Sm/cm at 1173 K was obtained for $Ce_{0.8}(Sm_{0.8}Sr_{0.2})_{0.2}O_{2-\delta}$.

It was shown that co-doping with SrO and BaO not only leads to the increase of electrical conductivity of solid electrolytes based on CeO$_2$, but also expands their electrolytic area. The lowest value $3.30\cdot10^{-23}$ atm of the critical oxygen partial pressure $p_{O_2}^*$ at which the values of the electronic and ionic components of conductivity are equal was found for $Ce_{0.8}(Sm_{0.75}Sr_{0.2}Ba_{0.05})_{0.2}O_{2-\delta}$. These materials can be recommended for use in intermediate temperature electrochemical devises including SOFCs.

REFERENCES

[1] V. V. Kharton, F. M. Figuiredo, L. Navarro, E. N. Naumovich, A. V. Kovalevsky, A. A. Yaremchenko, A. P. Viskup, A. Carneiro, F. M. B. Marques and J. Frade, Ceria -based materials for solid oxide fuel cells, *J. Mat. Scien.*. **36**. 1105-17 (2001).

[2] H. Yahiro, K. Eguchi and H. Arai, Electrical properties and reducibility of ceria-rare earth oxide systems and their application to solid oxide fuel cell, *S. S. I.*, **36**, 71-75 (1989).

[3] G.B. Balazs and R.S. Glass, AC-Impedance Studies of Rare-Earth-Oxide Doped Ceria, *S. S. I.*. **76**. 155-62 (1995).

[4] S. J. Hong and A. V. Virkar, Lattice-Parameters and Densities of Rare-Earth-Oxide Doped Ceria Electrolytes, *J. American Ceram. Soc.*, **78**, 433-39 (1995).

[5] S. Sameshima, H. Ono, K. Higashi, K. Sonoda, Y. Hirata and Y. Ikuma, Electrical Conductivity and Diffusion of Oxygen Ions in Rare-Earth-Doped Ceria, *J. Ceram. Soc. Jap.*, **108**, 1060-66 (2000).

[6] L. Minervini, M.O. Zacate and R.W. Grimes. Defect Cluster Formation in M_2O_3-Doped CeO$_2$, *S. S. I.* **116**, 339-49 (1999).

[7] H. Yoshida, T. Inagaku, K. Miura, M. Inaba and Z. Ogumi, Density functional theory calculation on the effect of local structure of doped ceria on ionic conductivity, *S. S. I.*, **160**, 109-16 (2003).

[8] F. Ye. T. Mori, D. R. Ou and A. N. Cormak, Dopant type dependency of domain development in rare-earth-doped ceria: An explanation by computer simulation of defect clusters, *S. S. I.*, **180**, 1127-32 (2009).

[9] H. Inaba and H. Tagawa, Ceria-based solid electrolytes, *S. S. I.*, **83**, 1-16 (1996).

[10] M. Mogensen, N. M. Sammes and G. A. Tompsett, Physical. chemical and electrochemical properties of pure and doped ceria, *S. S. I.*, **129**, 63-94 (2000).

[11] V. V. Kharton, A. A. Yaremchenko, A. A. Naumovich and F. M. B. Marques, Research on the electrochemistry of oxygen ion conductors in the former Soviet Union. III. HfO$_2$-. CeO$_2$- and ThO$_2$-based oxides, *J. S. S. Electrochem.*, **4**. 243-66 (2000).

[12]J. Van Herle, D. Seneviratne and A. J. McEvoy, Lanthanide. Co-doping of Solid Electrolytes: AC Conductivity Behaviour, *J. Europ. Ceram. Soc.*, **19**, 837-841 (1999).

[13]S. Omar, E. D. Wachsman and J. C. Nino, A co-doping approach towards enhanced ionic conductivity in fluorite-based electrolytes, *S. S. I.*, **177**, 3199-3203 (2006).

[14]M. Dudek, Ceramic oxide electrolytes based on CeO$_2$ – Praparation. properties and possibility of application to electrochemical devices, *J. Europ. Ceram. Soc.*, **28**, 965-71 (2008).

[15]H. Yamamura, E. Katoh, M. Ichikawa, K. Kakinuma, T. Mori and H. Haneda, Multiple Doping Effect on the Electrical Conductivity in the (Ce$_{1-x}$La$_x$M$_y$)O$_{2-\delta}$ (M = Ca. Sr), *Electrochem.*, **68**, 455-59 (2000).

[16]X. Sha, Z. Liu, X. Huang, J. Miao, L. Jia, X. Xin and W. Su, Preparation and properties of rare earth co-doped Ce$_{0.8}$Sm$_{0.2-x}$Y$_x$O$_{1.9}$ electrolyte materials for SOFC, *J. All. Comp.*, **424**, 315-21 (2006).

[17]S. Hui, J. Roller, S. Yick, X. Zhang, C. Decos-Petit, Y. Xie, R. Maric and D. Ghosh, A brief review of the ionic conductivity enhancement for selected oxide electrolytes, *J. Pow. S.*, **172**, 493-502 (2007).

[18] E.Yu. Pikalova, A.A. Murashkina, V.I. Maragou, A.K. Demin, V.N.Strekalovsky, P. Tsiakaras, Doped Ceria Systems as Promising Materials for Intermediate Temperature Electrochemical Devices

[19] E.Yu. Pikalova, V.I. Maragou, A.N. Demina, A.K. Demin and P.E. Tsiakaras, The effect of co-dopant addition on the properties of Ln$_{0.2}$Ce$_{0.8}$O$_{2-\delta}$ (Ln = Gd, Sm, La) solid-state electrolyte, *J. Power Sources*, **181**, 199-206 (2008).

[20] R. D. Shannon, Revised Effective Ionic Radii and Systamatic Studies of Interatomic Distances in Halides and Chalcogenides, *Acta Cryst.*, **A32**, P. 751 -67 (1976).

[21] E.Yu. Pikalova, A.V. Nikonov, B.D. Zhuravlev, V.G. Bamburov, O.M. Samatov, A.S. Lipilin, V.R. Khrustov, I.V. Nikolaenko, S.V. Plaksin, N.G. Molchanova, Effect of the Synthesis Technique on the Physicochemical Properties of Ce$_{0.8}$(Sm$_{0.75}$Sr$_{0.2}$Ba$_{0.05}$)$_{0.2}$O$_{2-\delta}$, *Inorg. mater.*, **47**, P. 396-401 (2011).

[22] T.-H. Yeh and C.-C. Chou, Ionic conductivity investigation in samarium and strontium co-doped ceria system, *Phys. Scripta*, **T129**, 303-07 (2007).

[23] H. Yahiro, T. Ohuchi, K. Eguchi and H. Arai, Electrical properties and microstructure in the system ceria-alkaline earth oxide, *J. Mat. Scien.*, **23**, 1036-41 (1988).

[24] E. G. Vaganov, V. P. Gorelov, N. M. Bogdanovich, I. V. Korzun and V. A. Kazantsev, Electroconduction and linear expansion of solid electrolytes Ce$_{1-x}$Sm$_x$O$_{2-\delta}$ (x = 0.10–0.30), *Rus. J. Electrochem.*, **43**. 663-66 (2007).

[25]K. Huang, M. Feng and J. B. Goodenough, Synthesis and electrical properties of dense Ce$_{0.9}$Gd$_{0.1}$O$_{1.95}$ ceramics, *J. Am. Ceram. Soc.*, **81**, 357-62 (1998).

[26] D.-J. Kim, Lattice Parameters. Ionic Conductivities. and Solubility Limits in Fluorite-Structure MO$_2$ Oxide (M=Hf^{4+}. Zr^{4+}. Ce^{4+}. Th^{4+}. U^{4+}) Solid Solutions, *J. Am. Ceram. Soc.*, **72**, 1415-21 (1989).

[27] D. Andersson. From the Electronic Structure of Point Defects to Functional Properties of Metals and Ceramics: Doctoral Thesis, Royal Institute of Technology (Sweden), Stockholm (2007).

[28] X. Sha, Z. Lü, X. Huang, J. Miao, L. Jia, X. Xin and W. Su, Preparation and properties of rare earth co-doped Ce$_{0.8}$Sm$_{0.2-x}$Y$_x$O$_{1.9}$ electrolyte materials for SOFC, *J. All. Comp.*, **424**, 315-21 (2006).

[29] P.-S. Cho, S. B. Lee, D.-S. Kim, J.-H. Lee, D.-Y. Kim and H.-M. Park. Improvement of Grain-Boundary Conduction in Gadolinia –Doped Ceria by Addition of CaO, *Electrochem. S. S. Let.*, **9**, A399-A402 (2006).

[30] B. S. Banerjee, P. S. Davi, P. Topwal, S. Mandal and K. Menon, Enhanced Ionic Conductivity in Ce$_{0.8}$Sm$_{0.2}$O$_{1.9}$: Unique Effect of Calcium Co-doping, *Adv. Func. Mater.*, **17**, 2847-54 (2007).

[31]A. Moure, J. Tartaj, C. Moure, Synthesis, sintering and electrical properties of gadolinia-calcia-doped ceria, *Mat. Let.*, **65**, 89-91 (2011).

[32]M. Dudek, A. Rapacz-Kmita, M. Mroczkowska, M. Mosialek and G. Mordarski, Co-doped ceria-based solid solution in the CeO$_2$-M$_2$O$_3$-CaO, M=Sm, Gd system, *Electrochem. Acta.*, **55**, 4387-94 (2010).

[33] T. Mori and H. Yamamura, Preparation of an Alkali-Element or Alkali-Earth-Element-Doped CeO$_2$– Sm$_2$O$_3$ System and Its Operation Properties as the Electrolyte in Planar Solid Oxide Fuel Cells, *J. Mat. Synth. Proc.*, **6**, 175-179 (1998).

Electrodes

ELECTROCHEMICAL PHENOMENA IN MEA ELECTRODES

Mihails Kusnezoff, Nikolai Trofimenko, Alexander Michaelis

Fraunhofer Institute for Ceramic Technologies and Systems
Winterbergstr. 28, 01277 Dresden, Germany

ABSTRACT

The numerous measurements of I-V-characteristics and impedance spectra on symmetrical cells and MEAs with different types of electrodes have been carried out at different temperatures using various gas compositions to characterize the performance of the electrolyte supported cells (ESC). The tests have been performed in ceramic housing in absence of any contamination sources resulting from experimental setup. The cathodic and anodic reaction steps have been investigated as a function of gas composition. It was found that the cathode polarization can be described by three processes: (i) oxygen adsorption on the cathode surface, (ii) oxygen diffusion to the three phase boundary followed by electrochemical reaction and (iii) oxygen transfer into electrolyte. In the MEA impedance spectra mainly the processes (i) and (ii) were observed. For B-doped LSM based cathodes the impact of oxygen adsorption on the total polarization resistance was reduced. In the anode the number of observed processes depends strongly on the art of the used anode and on the fuel humidification level. The Ni/YSZ anodes have mainly two arcs in impedance spectra connected with (i) hydrogen adsorption on Ni surface and (ii) diffusion of adsorbed species and formation of water. In the ceria based anodes the additional electrochemical reaction on ceria has been seen. With both anodes the ASR below 0.2 Ωcm^2@ 850°C for 10Sc1CeSZ based ESCs is achieved.

INTRODUCTION

For several years, the Fraunhofer Institute for Ceramic Technologies and Systems (IKTS) have been developing high performance electrolyte supported planar solid oxide fuel cells (ESC). The long-term, redox and thermal cycling stability of electrolyte supported cells based on 10Sc1CeSZ has been already achieved by optimizing of the electrodes. The degradation rate reported after 3,000 h of operation was lower then 0.1%/1000h. Developed cells satisfied all major requirements (thermal cycling, accident events such as load throw-off, abrupt cooling down and anode oxidation) for use in robust SOFC systems. Although significant gain of cell performance has been achieved in recent years, there is still considerable opportunity for further improvement through the optimization of materials and microstructure of electrodes combined with engineering focus on volume manufacturing processes (screen printing, co-firing etc.), reproducibility and cost reduction.

The total resistance of the membrane-electrode-assembly consists of the electrolyte resistance (ohmic loss arising from the ionic resistivity in the solid electrolyte) and the polarization resistances of the cathode and anode.

The cathodic polarization losses in the electrochemically active layer are defined by oxygen transport through the porous electrode and the oxygen reduction kinetics at the electrochemical active sites called "three phase boundaries" (or TPB) due to the simultaneous co-existence of three phases: electronic conductor and ionic conductor as a solid phases and molecular oxygen in the gas phase. In SOFC cathode oxygen from the air is reduced through the following reaction [1]:

$$O_2 + 2V_O^{..} + 4e^- \rightarrow 2O_O^x \tag{I}$$

The most common cathode materials used for SOFC are (La,Sr)MnO$_3$ (LSM) and (La,Ca)MnO$_3$ (LCM) perovskites. Due to their good catalytic activity for oxygen reduction and their chemical and thermo-mechanical compatibility to doped ZrO$_2$-electrolyte these materials can be directly applied as

SOFC cathodes in the temperature range of 800-1000°C. The polarization resistance reported for LSM-cathodes sintered at 1300°C is in the range of 1.5-19.8 Ωcm^2 at 950°C.

In the porous electrode the oxygen transport in the gas phase takes place in the pores of the cathode. Different paths were proposed for electrochemical oxygen transport in the cathode (Fig. 1). Generally accepted process of oxygen reduction can be described as follows: the oxygen is adsorbed on the surface of electronic conductor (i.e. LSM) and the adsorbed species diffuse to the interface electronic/ionic (i.e. LSM/YSZ) conductor. At this interface adsorbed oxygen species go into the electrolyte lattice as oxygen ions (Fig. 1). The rate determining steps of the overall reaction (I) depend on the cathode material and are still under discussion. In this paper we'll try to provide our view on this process.

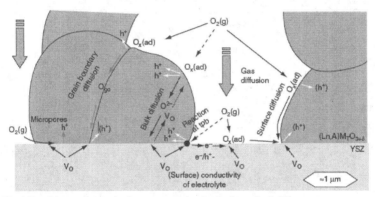

Fig. 1 Possible reaction paths for oxygen reduction in the cathode [2].

(La,Sr)MnO$_3$ perovskites have very low ionic conductivity in comparison with doped zirconia and the cathode containing mixture of LSM and YSZ allows to increase the three phase boundary length and the ionic conductivity of the cathode. In agreement to this it is found that the most efficient way to lower the polarization resistance of the cathode is to use a mixture of perovskite and electrolyte powders as an electrochemical active interlayer referred to as composite cathode. Other possibility, which was widely explored, was to utilize mixed ionic electronic conductors (i.e. La$_{0.6}$Sr$_{0.4}$Co$_{0.8}$Fe$_{0.2}$O$_3$) for direct surface oxidation and bulk transport of oxygen into electrolyte. Usually the cathodic polarisation resistance increases on decreasing the oxygen partial pressure in the oxidant and the operation temperature.

The anodic polarization resistance depends on the fuel gas composition. Using pre-converted natural gas as a fuel the following reactions take place in the anode:

1a Conversion:	$H_2 + O^{2-}_{el} - 2e^- \quad H_2O$	(II)
1b Conversion:	$CO + O^{2-}_{el} - 2e^- \quad CO_2$	(III)
2 Reforming:	$CH_4 + H_2O \quad CO + 3H_2$	(IV)
3 Water shift reaction:	$CO + H_2O \quad CO_2 + H_2$	(V)

In the presence of significant amounts of hydrogen (>20 Vol.%) and water vapor (>6 Vol%) in the fuel, the oxidation of hydrogen to water (1a) is the reaction most likely to happen in the anode. In this case instead of the electrochemical oxidation of carbon monoxide (1b) a water shift reaction (3) resulting in H_2 and CO_2 production in the gas phase takes place.

The main function of the anode is to promote the electrochemical oxidation of the fuel gases. The anode is also the layer of the MEA where the electrons have been released. Several materials were studied as SOFC anode materials such as noble metals ruthenium, rhodium, palladium, silver, platinum and gold, and as transition metals manganese, iron, cobalt, nickel and copper in the early stages of SOFC development. Platinum is a good catalyst however high vapour pressure of oxides at 900 °C – 1000 °C hinders the use in SOFC anodes. Gold shows no catalytic activity and poor adhesion to oxides. Ruthenium has high melting point (2310 °C) therefore low grain growth during operation and high catalytic activity for steam reforming and show negligible carbon deposition; however evaporation of ruthenium oxide above 1200 °C is a problematic issue [3]. Besides these factors the high costs of noble materials is a negative aspect to utilize them in the anode. Among the transition metals nickel is a typical material used in solid oxide fuel cells. Nickel is cheap and has been shown as having the highest catalytic activity among different anode materials in a systematic study [4]. It has high chemical stability and good chemical compatibility with other SOFC components [5] and low cost. Among these advantages, nickel also has some drawbacks. It expands more than 30% in oxidising atmosphere which causes delamination of the anode after redox cycles. Under high partial pressure of water $Ni(OH)_2$ species can be formed which cause long term degradation [6]. Finally nickel has a high sensitivity towards sulfur and carbon containing gases in terms of deactivation resulting in performance loss and degradation of anode.

Nickel is usually combined with ceramic material. The use of yttria stabilised zirconia (YSZ) with nickel as SOFC anode by Spacil in 1970s showed promising results and since then nickel is most widely used SOFC anode material [7]. YSZ phase in the anode provides the oxygen ion conductivity and nickel phase the electronic conductivity (Fig. 2). Mixing YSZ to nickel lowers the thermal expansion coefficient and reduces nickel agglomeration in the anode.

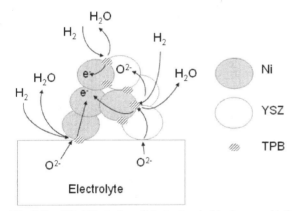

Fig. 2 Simplified illustration of electrochemical hydrogen oxidation reaction at triple phase boundaries (TPB) of a Ni/YSZ anode.

There are several models for the kinetics for the adsorption, surface and interface reactions in the anode. Bieberle [6] predicted the model shown in Fig. 3 for the kinetics of the SOFC anodes. According to Bieberle, the electrochemical reactions take place only on Ni surface and TPB; however it is also assumed that on YSZ surface the formation of hydroxyl and chemisorption of water are also possible. It is generally accepted that H_2 adsorbs dissociatively on Ni. The charge transfer reaction is assumed to take place in one step [6]. Further reaction doesn't provide charge transfer and give some response in the impedance spectra only due to diffusion processes coupled to this charge transfer reaction.

A common praxis in the literature is to model electrochemical processes by equivalent circuits that contain a sum of linear elements describing the major processes (i.e. ohmic loss, interfacial charge transfer, concentration polarization). The noncharge-transfer processes (solid state diffusion, surface exchange and gas-phase diffusion) can not be resolved individually into additive terms contributing to the total cell impedance. Nevertheless the "electrochemical" diffusion related elements such as Warburg impedance and Gerisher impedance can be successfully used to fit the impedance spectra and to understand the rate determining steps of reaction.

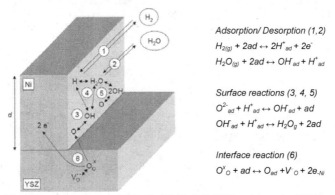

Adsorption/ Desorption (1,2)

$H_{2(g)} + 2ad \leftrightarrow 2H^+_{ad} + 2e^-$

$H_2O_{(g)} + 2ad \leftrightarrow OH_{ad} + H^+_{ad}$

Surface reactions (3, 4, 5)

$O^{2-}_{ad} + H^+_{ad} \leftrightarrow OH_{ad} + ad$

$OH_{ad} + H^+_{ad} \leftrightarrow H_2O_g + 2ad$

Interface reaction (6)

$O^x_O + ad \leftrightarrow O_{ad} + V^{..}_O + 2e_{-Ni}$

Fig. 3 Electrochemical model for the Ni, H_2-H_2O/YSZ system under SOFC operating conditions [6].

To separate the impact of the different contributions to the electrochemical resistance, impedance spectroscopy is used. One of latest interpretation of MEA spectra available in literature is shown in Fig. 4. According to the figure, the high frequency region of Nyquist plot represents the gas diffusion in the anode coupled with charge transfer reactions whereas low frequencies represents the O_2- diffusivity and surface exchange kinetics in the cathode as well as mass transfer resistance in the anode.

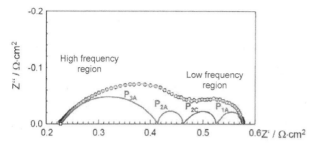

Fig. 4 Nyquist plot of the impedance data in and corresponding semicircles of various processes. (P1A: mass transfer resistance in the anode, P2A, P3A: gas diffusion coupled with charge transfer reaction and ionic transport (at anode functional layer), P1C: gas diffusion (cathode – not shown due to low values), P2C: oxygen surface exchange kinetics and O_2- diffusivity in the cathode) (Modified from [8]).

In our paper we'll discuss the origin and the contribution of different processes to impedance spectra and especially to the polarization resistance.

PROCESSES AND EQUIVALENT CIRCUITS FOR EXPLANATION OF IMPEDANCE SPECTRA

Cathodic reaction

The kinetics of oxygen reduction is widely discussed in the literature but still is not fully understood in detail. On basis of literature data and own experience we suppose following mechanism (see also Fig. 5) for oxygen reduction (Eq. II):

$$O_2 \text{ (gas)} + (s) = h(s) + (s\text{-}O_2)^{\cdot} \text{ (}1^{st} \text{ rate limiting step)} \tag{I.1}$$

$$(s\text{-}O_2)^{\cdot} + (s_{TBP}) = h \text{ (e.c.)} + 2 \text{ }(s_{TBP}\text{-}O)^{\cdot} \tag{I.2}$$

$$(s_{TPB}\text{-}O)^{\cdot} + V\ddot{o}(s_{TBP}) = h \text{ (e.c.)} + (s_{TPB}) \text{ (}2^{nd} \text{ rate limiting step)} \tag{I.3}$$

$$V\ddot{o}(s_{TPB}) = V\ddot{o} \text{ (i.c.)} \tag{I.4}$$

where (s) is a site on the surface of electronic conductor (e.c.) or mixed conductor (m.c.) and (s_{TPB}) is the surface of TPB, $(s\text{-}O_2)^{\cdot}$ is the adsorbed oxygen molecule and $(s\text{-}O)^{\cdot}$ is an adsorbed oxygen atom. We suppose that the reaction on the surface of the electronic conductor (I.1) corresponds to the "classical" surface exchange rate of active electrode material. We assume that the adsorbed oxygen molecules can be much more easily converted to adsorbed oxygen atoms at the electronic conductor / ionic conductor interface due to enhanced concentration of vacancies in this region. Therefore adsorbed oxygen molecules diffuses to TPB due to concentration gradient of adsorbed oxygen species on the surface of electronic conductor. In case of mixed ionic / electronic conductor as electrode material the reactions (I.1 – I.3) can take place directly on the surface of electrode as claimed by Adler et al. [1] and at high PO_2 the reaction (I.1) is not more rate limiting. However at low oxygen partial pressures the reaction (I.1) becomes again the rate limiting step.

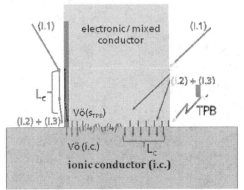

Fig. 5 Proposed mechanism of cathodic reaction for electronic (blue lines) and mixed conductor (red lines).

The impedance reaction (I.1) can be derived from appropriate expression for interfacial charge transfer reaction with a single activated rate limiting step:

$$i_{I.1} = i_{0_{I.1}} \left[\exp\left(\frac{\alpha_{c,I.1} F}{RT} \eta_{I.1} \right) - \exp\left(-\frac{\alpha_{a,I.1} F}{RT} \eta_{I.1} \right) \right] + C_{I.1} \frac{d\eta_{I.1}}{dt} \tag{1}$$

where $C_{I.1}$ is the surface capacitance of the interface, $i_{0_{I.1}}$ is the exchange current density between atmosphere and surface of electronic conductor and $\alpha_{c,I.1}$, $\alpha_{a,I.1}$ are the Tafel constants for cathodic and anodic reactions.

The resulting impedance of process (II.1) can be easily obtained inserting $\eta_{I.1} = \eta_o e^{j\omega t}$ is

$$\frac{1}{Z_{I.1}} = \frac{i_{0_{I.1}} (\alpha_{c,I.1} - \alpha_{a,I.1}) F}{RT} (1 - j\omega\tau_{I.1}) \tag{2}$$

with $\tau_{I.1} = \dfrac{C_{I.1} RT}{i_{0_{I.1}} (\alpha_{c,I.1} + \alpha_{a,I.1}) F}$.

The derived impedance expression corresponds to the simple parallel RC-circuit and delivers an arc in Nyquist impedance plot.

We suppose that $C_{I.1}$ is the capacitance of the entire catalyst surface (similar to pseudocapacitance of supercap), because in this case the charge accumulation by oxygen exchange between surface of catalyst and gas phase takes place.

General relationship between current density and interface capacity can be written as

$$i = C_{I.1} \frac{d\eta_{I.1}}{dt} = zF \frac{\Gamma d\theta_{O_2^-}}{dt} \frac{S}{A} \tag{3}$$

with Γ – total number of the adsorption sites per unit area, $\theta_{O_2^-}$ – steady state degree of coverage of the adsorbates, S – free surface area of catalyst in the printed electrode and A – electrode area.

If we assume the Temkin isotherm for oxygen adsorption / desoption

$$\frac{P_{O_2}}{P_{tot}} = \exp(\Delta H^0_{O_2}(1-\alpha\theta_{O_2^-})/RT - \Delta S^0_{O_2}/R) \qquad (4)$$

with P_{O_2} – oxygen partial pressure, P_{tot} – total gas pressure, $\Delta H^0_{O_2}$ – enthalpy of oxygen adsorption, $\Delta S^0_{O_2}$ – entropy of oxygen adsorption, α – constant;
than we find following relation between overpotential and steady state coverage degree of adsorbed oxygen :

$$d\theta_{O_2^-} = RT\frac{d\ln\dfrac{P_{O_2}}{P_{tot}}}{\alpha\Delta H^0_{O_2}} = zF\frac{d\eta_{I.1}}{\alpha\Delta H^0_{O_2}} \qquad (5)$$

The interface capacitance can be easy obtained inserting Eq. (5) in Eq. (3) and taking into account the surface area of catalyst per geometric area of the electrode:

$$C_{I.1} = \frac{(zF)^2\Gamma}{\alpha\Delta H^0_{O_2}(T)}\cdot\frac{S}{A} \qquad (6)$$

The absolute value of area specific capacitance has been estimated from Eq. (6) using following values: $\Gamma = 10^{-9}$ mol/cm², $\alpha = 0.1$, $\Delta H^0_{O_2} = 10$ kJ/mol, S $= 0{,}036 - 0{,}072$ m²/cm². The estimation shows that the capacitance has to be as high as $100..1000$ mF/cm². We suppose that $\Delta H^0_{O_2}$ doesn't change very much in temperature range 700-850°C and therefore the capacity is almost independent on the temperature. The polarization resistance of the interface is reversibly proportional to the exchange current density at the catalyst surface $i_{0_{I.1}}$:

$$R_{I.1} = \frac{RT}{i_{0_{I.1}}(\alpha_{c,I.1}+\alpha_{a,I.1})F} \qquad (7)$$

The exchange current density at the catalyst surface can be calculated using following expression often used in the litearture [9,10] to estimate the exchange current density at TPB assuming surface exchange as rate controlling step:

$$i_{0_{I.1}} = \frac{F}{V_m}\cdot K \qquad (8)$$

here K – is the surface exchange rate of oxygen between catalyst surface and gas phase and V_m the molar volume of mixed oxide. The surface exchange coefficient K is proportional to the reaction constant of Eq. I.1 .

The exchange current density dependence on partial pressure can be simply estimated from the first principles of electrochemistry [11]:

$$i_{0_{I.1}} \approx const \cdot \left(\frac{P_{O_2}}{P_{O_2}^0} \right)^{\alpha_{c,I.1}} \tag{9}$$

with $\alpha_{c,I.1} = n \cdot \gamma$

where n is a number of transferred electrons during the rate determining step (rds) and is the symmetry of rds. In case of symmetrical reaction $\gamma=0,5$ and one transferred electron in reaction I.1 one get $|Z_{I.1}| = R_{I.1} \approx const \cdot P_{O_2}^{-0,5}$. The temperature dependence of $R_{I.1}$ as well as $i_{0_{I.1}}$ should follow Arrenius plot (see Eq. 1).

Reactions (I.2) and (I.3) are related to each other by diffusion of oxygen species (independently whether the diffusion on the surface (electronic conductor as catalytic active material) or in the bulk (mixed conductor as catalytic active material) takes place). In this case the impedance should be calculated using appropriate continuum equations for diffusion of adsorbed oxygen species on the surface of electronic conductor (or in the bulk of mixed conductor) and suitable boundary conditions. The mathematical derivation of impedance equations was made assuming that the process I.1 is not the rate limiting step. The more detailed information on this can be found in [1] and [12].

In case of mixed ionic / electronic conductor one get following expression

$$Z_{I.2} = \frac{RT}{(1-p)(zF)^2 \, \tilde{D}_V \, c_{V\delta}^0} \left(\frac{\tilde{D}_V \, \tau_{I.2}}{1 + j\omega\tau_{I.2}} \right)^{1/2} \tag{10}$$

with $\tilde{D}_V$ – chemical diffusion coefficient of the oxygen vacancies

       $c_{V\delta}^0$ – the concentration of the oxygen vacancies in mixed conductor

       p – porosity of the cathode

       $\tau_{I.2} = d_0/4\,\beta$ – characteristic time for oxygen transfer into perovskite grain with characteristic particle size d_0.

       β – exchange coefficient at TBP, which is proportional to the reaction constants of reactions I.1-I.3 and implicitly depends from oxygen exchange reaction on the surface of catalyst.

The polarization resistance in case of mixed ionic/electronic conductor depends on both surface exchange and ionic conductivity according following equation:

$$R_{I.2} = \frac{RT}{(1-p)(zF)^2 \, \tilde{D}_V \, c_{V\delta}^0} \left(\tilde{D}_V \, \tau_{I.2} \right)^{1/2} = \frac{L_c}{(1-p)\sigma_V} \tag{11}$$

with $\sigma_V = \dfrac{c_{V\delta}^0 \, \tilde{D}_V (zF)^2}{RT} = \left[c_{V\delta}^0 = \dfrac{\delta}{V_m} \right] = \dfrac{\delta \tilde{D}_V (zF)^2}{V_m RT}$ $\qquad$ (12)

where δ is the ☐ non-stoichiometry and V_m the molar volume of the mixed conductor. D_V is almost independent on PO_2. The PO_2 dependence of $R_{I.2}$ follows the PO_2 dependence of term $\left(c_{V_0}^0 \beta^{1/2}\right)^{-1}$. If we assume that $c_{V_0}^0 \approx const \cdot P_{O_2}^{-n_1}$ and $\beta \approx const \cdot P_{O_2}^{n_2}$ then the dependence of $\log R_{I.2}$ on $\log PO_2$ can be calculated as:

$$n_R = n_1 - n_2/2 \tag{13}$$

The dependence of $c_{V_0}^0$ and β on PO_2 is for perovskite material of opposite sign and results in weaker (or no) dependence of $R_{I.2}$ from PO_2 compared to $R_{I.1}$.
$R_{I.2}$ values estimated for pure LSCF, LSC and LCCF electrodes [1] are in the range of 0,2-10 Ωcm^2.
The dependence of $R_{I.2}$ on temperature follow the Arrenius plot with activation energy which is a product of activation energies for ionic conductivity of m.c. and of oxygen rate exchange at TPB.
For this process also the value of electrochemical capacitance can be calculated:

$$C_{I.2} = \frac{(zF)^2 c_{V_0}^0}{RT}(\tilde{D}_V \tau_{I.2})^{1/2} \cdot (1-p) \cdot A = \frac{(zF)^2 c_{V_0}^0 L_c A}{RT} = \frac{\sigma_V L_c (1-p) A}{D_V} \tag{14}$$

where L_c is the electrochemically active thickness of the electrode (extension of TPB) and A is the area of the electrode. For typical mixed conductor materials the area specific capacitance values has been estimated to be in order of 10..1000 mF/cm^2. Capacitance dependence on PO_2 can be calculated from $\left(c_{V_0}^0 \beta^{-1/2}\right)$ as

$$n_C = -n_1 - n_2/2 \tag{15}$$

Thus estimating n_R and n_C from impedance spectra the dependence of $c_{V_0}^0$ and β from PO_2 can be calculated. From Eq. (14) capacitance dependence on temperature can not be easily estimated.
In case of pure electronic conductor we get similar expressions but with different physical meaning. The derivation of these equations is described in [12]. For impedance we obtain following expression:

$$Z_{I.2} = \frac{1}{TPBL}\frac{RT}{\Gamma(zF)^2 \theta_{eq}^0 D_\theta}\left(\frac{D_\theta}{K}\right)^{1/2}\frac{1}{(1+j\omega\tau_{I.2})} \tag{16}$$

with Γ – number of the adsorption sites per unit area
θ^0_{eq} – steady state degree of coverage of the adsorbates
D – surface diffusion coefficient of the adsorbed oxygen species
K – surface exchange coefficient (proportional to the reaction constant for I.3)
TPBL – three phase boundary length per unit electrode area
$\tau_{I.2} = 1/K$ – characteristic relaxation time.

The polarization resistance depends in this case on TPBL, surface diffusion coefficient, surface concentration of adsorbed species ($\Gamma \theta^0_{eq}$) and surface reaction coefficient K

$$R_{I.2} = \frac{1}{TPBL}\frac{RT}{\Gamma(zF)^2 \theta_{eq}^0 D_\theta}\left(\frac{D_\theta}{K}\right)^{1/2} = \frac{1}{TPBL}\frac{RT}{\Gamma(zF)^2 \theta_{eq}^0 D_\theta}L_c \tag{17}$$

The Po$_2$ dependence of the polarization resistance is mainly defined by term $K^{-1/2}$. Compared to $R_{I.1}$ which is proportional to K^{-1} the $R_{I.2}$ should have two times weaker dependence on log PO$_2$. The electrochemical capacitance can be calculated as

$$C_{I.2} = \frac{(zF)^2 \Gamma \theta_{eq}^0}{RT}(D_\theta \tau_{I.2})^{1/2} \cdot TPBL = \frac{(zF)^2 \Gamma \theta_{eq}^0 L_C}{RT} \cdot TPBL \qquad (18)$$

where L_c is the electrochemically active thickness of the electrode (extension of TPB).

The temperature dependence of the resistance and capacitance should obey the Arrenius equation and increase with the temperature. If we assume that θ_{eq}^0/T is constant (see Eq. (5)) than we get that resistance and capacitance have similar dependence on PO$_2$ ($R_{I.2} \sim K^{-1/2}$ and $C_{I.2} \sim K^{-1/2}$) and increases decreasing oxygen partial pressure. In case where θ_{eq}^0 or Γ is proportional to the vacancy concentration in the perovskite the PO$_2$ dependence can be still described by equations (13) and (15).

The dependence on temperature is defined by term $(D_\theta \tau_{I.2})^{1/2} = (D_\theta / K)^{1/2}$. Because the activation energies for diffusion and surface exchange are of the same order of magnitude the capacitance $C_{I.2}$ would increase with temperature in case of higher activation energy for surface exchange rate in comparison with activation energy for diffusion and decrease in case of lower activation energy for surface oxygen exchange.

It was generally found that only for selected materials the bulk diffusion mechanism (Eq. (5)) can describe the phenomena in the air electrode. The most reasonable mechanism for common SOFC cathodes (i.e. La$_{1-x}$Sr$_x$MnO$_3$) is the surface diffusion with following charge transfer reaction at TPB. The Nyquist plot for the described reaction mechanisms is shown in Fig. 6 and is often referred as Gerisher impedance.

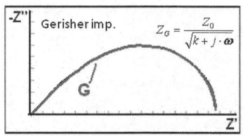

Fig. 6 Appearence of Gerisher impedance in Nyquist plot.

The interfacial charge transfer reaction (I.4) describes the transfer of vacancies from TPB across the electronic (or mixed) conductor / ionic conductor interface. Here we can use a rate expression appropriate for an interfacial charge transfer reaction with a single activated rate-limiting step:

$$i_{I.4} = i_{0_{I.4}}\left[\exp\left(\frac{\alpha_{c,I.4}F}{RT}\eta_{I.4}\right) - \exp\left(-\frac{\alpha_{a,I.4}F}{RT}\eta_{I.4}\right)\right] + C_{I.4}\frac{d\eta_{I.4}}{dt} \qquad (19)$$

where $C_{I.4}$ is the surface capacitance of the electronic / ionic conductor interface, $i_{0_{I.4}}$ is the exchange current at this interface and $a_{c,I.1}$, $a_{a,I.1}$ are the Tafel constants for forward and backward vacancy jumps.

The resulting impedance of process (I.4) is

$$\frac{1}{Z_{I.4}} = \frac{i_{0_{I.4}}(a_{c,I.4} + a_{a,I.4})F}{2RT}(1 - j\omega\tau_{I.4}) \tag{20}$$

$$\text{with } \tau_{I.4} = \frac{C_{I.4}RT}{i_{0_{I.4}}(\alpha_{c,I.4} + \alpha_{a,I.4})F}. \tag{21}$$

We suppose that $C_{I.4}$ is the capacitance of the typical double layer at the ionic/electronic blocking interface and is typically in the range of 10-1000 $\mu F/cm^2$. The activation energy of polarization resistance for this process should be in the range of activation energy for ionic conductivity of electrolyte or mixed ionic conductor. We suppose only weak PO_2 dependence for this process. We want to point out that the absolute value of polarization resistance in this case would depend strongly on TPBL and isolating layers between electronic conductor and electrolyte and thus is very sensitive to $La_2Zr_2O_7$ or $SrZrO_3$ phase formation between perovskite and electrolyte well known from the literature. Without zirconate formation the interfacial resistance perovskite / electrolyte is supposed to be very small (negligible).

Anodic reaction on Ni/YSZ

The anodic reaction on Ni/YSZ is extensively discussed in [6]. Here we apply simplified approach for derivation of expressions for electrochemical impedance for Ni/YSZ anode. We suppose that two main electrochemical reactions take place in Ni/YSZ electrode with well dispersed Ni particles:

$$H_{2(g)} + 2ad \leftrightarrow 2H^+_{ad} + 2e^- \tag{II.1}$$

$$O^x_O + ad \leftrightarrow O_{ad} + V^{\cdot\cdot}_O + 2e^-_{Ni} \tag{II.2}$$

As agreed in the literature the dissociative hydrogen adsorption takes place in one step and we can treat the charge transfer reaction II.1 similar to the reaction I.1 and get following expressions for impedance and their constituents:

$$\frac{1}{Z_{II.1}} = \frac{i_{0_{II.1}}(a_{c,II.1} + a_{a,II.1})zF}{RT}(1 - j\omega\tau_{II.1}) \tag{22}$$

where $C_{II.1}$ is the surface capacitance of the nickel / fuel gas interface, $i_{0_{II.1}}$ is the exchange current density between gas phase and surface of nickel and $a_{c,II.1}$, $a_{a,II.1}$ are the Tafel constants for cathodic and anodic reactions.

We suppose that $C_{II.1}$ is the surface capacitance of the catalyst, because in this case the hydrogen exchange on nickel surface takes place. In this case the specific area capacitance can be calculated as follows:

$$C_{II.1} = \frac{(zF)^2 \Gamma}{\alpha \Delta H^0_{H_2}} \cdot \frac{S_{Ni}}{A}$$
(23)

with Temkin adsorption isotherm:

$$\frac{P_{H_2}}{P_{H_2O}} = \exp(\Delta H^0_{H_2}(1 - \alpha\theta_{H^+})/RT - \Delta S^0_{H^+}/R)$$

Thus $C_{II.1}$ has electrochemical nature it can reach relatively large absolute values. Due to this reason 1/RC-constant is rather high and the hydrogen adsorption can be found at low frequency region of impedance spectra. The semicircle of this reaction can overlap with that of reaction I.1 and therefore can be difficult resolved from impedance spectra of the cell.

The interface reaction (II.2) at Ni/YSZ interface has the same origin as cathodic reaction (I.4) and can be also described by simple RC-circuit (see Eq. 19-21). However the resistance of this reaction should be higher compared to air electrode due to pure metallic conductivity of nickel. The interfacial capacity in the range of 10-1000 μF/cm^2 is awaited for this process. PH$_2$ dependence for R$_{II.2}$ can not be easily predicted from the theoretical point of view.

<u>Gas diffusion polarisation</u>

Often the gas diffusion polarization caused by change of Nernst voltage due to gas concentration change in the pores of electrode can be observed in the impedance spectra. The gas diffusion polarisation resistance of the air electrode can be calculated using following equations and boundary conditions at the interface air electrode electrolyte:

$$-\frac{i}{zF} = -D_{12}\frac{dc_{O_2}}{dx} + vc_{O_2}$$
(24)

$$0 = -D_{21}\frac{dc_{N_2}}{dx} + vc_{N_2}$$
(25)

$$c_{O_2}\big|_{x=d} = c^*_{O_2}$$
(26)

$$c_{N_2}\big|_{x=d} = c^*_{N_2}$$
(27)

$$c_{O_2}(x) + c_{N_2}(x) = c_{tot} = \frac{P_{tot}}{RT} = const$$
(28)

where c_{O_2} is the oxygen concentration in the gas phase, D_{12} is the oxygen diffusion coefficient in the binary mixture N$_2$/O$_2$, c_{N_2} is the nitrogen concentration in the gas phase, D_{12} is the oxygen diffusion coefficient in binary mixture N$_2$/O$_2$, d – is the thickness of the electrode, x is the axis perpendicular to the electrolyte surface, $x = 0$ – is the interface electrode electrolyte and v is volume flow rate of gas towards the interface due to volume change by oxygen pumping through the cell.
The solution of differential equations (24), (25) with boundary conditions (26) and (27) can be simply derived assuming that $D_{12}=D_{21}$ and delivers following expression for $c_{O_2}(x)$:

$$c_{O_2}(x) = c_{tot} - (c_{tot} - c^*_{O_2})e^{\frac{i}{zFD_{21}c_{tot}}(d-x)}$$
(29)

The change of open circuit potential caused by oxygen pumping through the cell can be found from Nernst equation:

$$\Delta E_{OCV} = \frac{RT}{zF} \ln\left[\frac{c_{O_2}^*}{c_{O_2}(x=o)}\right] = \frac{RT}{zF} \ln\left[\frac{c_{O_2}^*}{c_{tot}(1-(1-\frac{c_{O_2}^*}{c_{tot}})e^{\frac{i}{zFD_{21}c_{tot}}d})}\right] = \left[X_{O_2} = \frac{c_{O_2}^*}{c_{tot}}\right]$$

$$= \frac{RT}{zF} \ln\left[\frac{1}{1-(1-X_{O_2})e^{\frac{i}{zFD_{21}c_{tot}}d}}\right]$$

$$(30)$$

For very small current ($i \to 0$) expression (30) can be simplified to

$$\Delta E_{OCV} = \frac{RT}{zF} \cdot \frac{i}{zFD_{21}c_{tot}} d \cdot \frac{1-X_{O_2}^*}{X_{O_2}^*} \tag{31}$$

From eq. (31) the polarization resistance can be easily estimated:

$$R_{I,Nernst} = \frac{\Delta E_{OCV}}{i} = \left(\frac{RT}{zF}\right)^2 \cdot \frac{1}{PD_{21}c_{tot}} d \cdot \frac{1-X_{O_2}^*}{X_{O_2}^*} \tag{32}$$

The capacity of gas volume can be also easily derived from general expressions:

$$i = C\frac{dE_{OCV}}{dt} = \frac{dc_{O_2}}{dt} \tag{33}$$

The oxygen concentration in the gas phase can be obtained from:

$$c_{O_2} = \frac{P_{O_2}}{RT} \tag{34}$$

and dE_{OCV} from Nernst equation as

$$dE_{OCV} = \frac{RT}{zF} \cdot \frac{dc_{O_2}}{c_{O_2}} \tag{35}$$

Inserting (34) and (35) in eq. (33) one can get

$$C_{I,Gas} = \left(\frac{zF}{RT}\right)^2 \cdot P_{O_2}V \tag{36}$$

The gas volume of electrode without current collector is typically 0,05 x 40 x 40 mm³ (80 mm³). Including current collector mesh as a part of electrode we get typical volume of 0,5 x 40 x 40 mm³ (800 mm³) for capacity estimation. The capacity is proportional to P_{O_2} and reverse proportional to square of the temperature.

The polarization resistances and capacitances of air electrode as a result of oxygen utilization at the TPB for different P_{O_2} at 850℃ are summarized in Table I (we have used effective value of $D_{O2}=\tau D_{21}=2,1\times10^{-4}$ m² /s with tortuosity τ factor of 0,5 for our estimations).

Table I   Polarization resistance and capacity values resulting from air conversion at 850℃.

	$X^{\cdot}_{O_2}=0,02$	$X^{\cdot}_{O_2}=0,04$	$X^{\cdot}_{O_2}=0,08$	$X^{\cdot}_{O_2}=0,21$
$R_{1,Nernst}$ in mΩccm²	6,46	3,16	1,52	0,50
$C_{1,Gas}$ in F	2,76	5,52	11,03	28,96

From these estimations one can see that the polarization resistance due to oxygen conversion is negligible (<1 mΩcm²) up to P_{O_2} =8%. The gas capacity of air electrode decreases reducing P_{O_2} and lies between 2 and 29 F in air at 850℃. Due to very small absolute values the oxygen diffusion polarization resistance can be neglected in case of ambient oxygen partial pressures and low oxygen utilization over the electrode (high air flow rates).

Similar to the cathodic conversion impedance the anodic conversion polarization resistance (referred also as gas diffusion resistance) can be found. In the case of anodic gas conversion there is no significant volume flow of gas towards the electrode / electrolyte interface because the conversion of H_2 to H_2O doesn't produce any molar change in the fuel. We consider following simplified equations for calculating the H_2O-concentration in gas phase at TPB due to hydrogen oxidation:

$$\frac{i}{zF} = -D'_{12}\frac{dc_{H_2O}}{dx} \tag{37}$$

$$-\frac{i}{zF} = -D'_{21}\frac{dc_{H_2}}{dx} \tag{38}$$

$$c_{H_2O}\big|_{x=d} = c^{\cdot}_{H_2O} \tag{39}$$

$$c_{H_2}\big|_{x=d} = c^{\cdot}_{H_2} \tag{40}$$

$$c_{H_2O}(x) + c_{H_2}(x) = c_{tot} = \frac{P_{tot}}{RT} = const \tag{41}$$

where c_{H_2O} is the vapour concentration in the gas phase, D'_{12} is the water gas diffusion coefficient in binary mixture H_2/H_2O, c_{H_2} is the hydrogen concentration in the gas phase, D'_{21} is the hydrogen diffusion coefficient in binary mixture H_2/H_2O, d – is the thickness of the electrode, x is the axis perpendicular to the electrolyte surface, $x = 0$ – is the interface electrode / electrolyte.
The hydrogen transport and water removal from the interface take place by gas interdiffusion. Generally we assumed that $D'_{12} = D'_{21}$.
The solution of differential equation (37) gives the water vapour concentration as a function of current through the interface:

$$c_{H_2O}(x) = c^{\cdot}_{H_2O} + \frac{i}{zFD'_{21}c_{tot}}(d-x) \tag{42}$$

The change of open circuit potential caused by water production at the interface can be found from Nernst equation:

$$\Delta E_{OCV} = \frac{RT}{zF} \ln \left[\frac{c^*_{H_2O}}{c_{H_2O}(x=o)} \cdot \frac{1 - c_{H_2O}(x=o)}{1 - c^*_{H_2O}} \right] \tag{43}$$

For small changes of H_2O concentration (ΔX_{H2O}) (small current induced during impedance measurement) due to water production on the anode following relation between water concentration and Nernst voltage can be found:

$$\Delta E_{OCV} = \frac{RT}{zF} \left[\frac{1}{X^*_{H_2O}} + \frac{1}{1 - X^*_{H_2O}} \right] \cdot \frac{i}{zFD'_{21} c_{tot}} d \tag{44}$$

From eq. (44) the polarization resistance due to water conversion can easily be derived

$$R_{II,Nernst} = \left(\frac{RT}{zF} \right)^2 \left[\frac{1}{X^*_{H_2O}} + \frac{1}{1 - X^*_{H_2O}} \right] \cdot \frac{d}{D'_{21} P} \tag{45}$$

Similarily to Eq. (25)-(29) one can obtain the gas capacity of fuel compartment:

$$C_{II,Gas} = \left(\frac{zF}{RT} \right)^2 \left[\frac{1}{X^*_{H_2O}} + \frac{1}{1 - X^*_{H_2O}} \right]^{-1} \cdot PV = \left(\frac{zF}{RT} \right)^2 X^*_{H_2O} \left(1 - X^*_{H_2O} \right) \cdot PV \tag{46}$$

The fuel conversion polarization resistance has minimum and gas capacity has maximum at 50:50 water / hydrogen gas mixture as fuel. Therefore using highly moistured hydrogen the anode conversion impedance can be experimentally eliminated.

Table II Polarization resistance and capacity values resulting from fuel conversion at 850°C.

	$X^*_{H_2O}=0.02$	$X^*_{H_2O}=0.05$	$X^*_{H_2O}=0.1$	$X^*_{H_2O}=0.5$
$R_{II,Nernst}$ in mΩccm²	13,45	5,55	2,93	1,05
$C_{II,Gas}$ in F	2,70	6,55	12,41	34,48

Table II shows the values of gas capacitance and gas diffusion polarization resistance of the anode as a function of $X^*_{H_2O}$ at 850°C (we have used effective value of $D'_{H2O}=\tau D'_{21}=8,4\times10^{-4}$ m²/s with tortuosity τ factor of 0,25 for our estimation). The values of polarization resistance have significant impact on overall losses only at $X^*_{H_2O} \leq 0,05$.

The gas conversion impedance appears in the impedance spectra at OCV as semicircle. During the current flow through the cell some suppression of semicircle takes place due to variation of polarization resistance and gas capacitance along the cell area due to different steam content in the fuel caused by current flow.

Equivalent circuit for deconvolution of impedance spectra

The equivalent circuit of a solid oxide fuel cell can be described as a combination of inductors, capacitors and resistors. We used following equivalent circuit for fitting the impedance spectra of MEA:

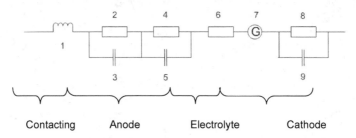

| Contacting | Anode | Electrolyte | Cathode |

Figure 8. Equivalent circuit for calculation of MEA impedance.

Basically, the circuit consists of one inductor for the contacting of the cell (1), three charge capacitance elements (3, 5, 9) and four resistance elements (2, 4, 6, 8). Elements 2 and 3 represent the process II.2, the 4th and 5th elements represent the process II.1. Element 6 represents resistance due to the ionic conductivity of the electrolyte which should follow the Ohm's law. 8th and 9th elements represent the adsorption of oxygen (process I.1), the 7th element is Gerisher element (Z_G) and represents a diffusion/charge transfer at TPB (process I.2 and I.3).
For symmetrical cell the equivalent circuit is described as follows (see Fig. 9):

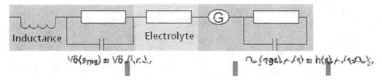

Figure 9. Equivalent circuit for calculation of impedance of symmetrical cell.

METHODS & EXPERIMENTALS
The lanthanum strontium manganite without ($La_{1-x}Sr_xMnO_{3-\delta}$, x=0.25) and with additional transition metal on B-place $La_{1-x}Sr_xMn_{1-y}M'_yO_{3-\delta}$ (LSMM′) , 8YSZ, Sc_2O_3 stabilized ZrO_2 (ScSZ), gadolinia doped ceria (GDC) and NiO powders used in this work were supplied by different manufacturers, according our specifications for the stoichiometry, crystalline phase, specific surface and particle size distributions.
The anode and cathode pastes were prepared using equipment and steps described in [13-15], which have been significantly modified and evaluated for up-scaling. Using developed paste processing routes, all pastes have been prepared with batch sizes up to 2 kg. Developed paste manufacturing method with optimized parameters allowed to produce pastes with desired properties (viscosity, solid

phase content, long-term stability etc.) independent from scattering of properties of raw materials. Additional wetting agents were used to improve screen printing characteristics such as thickness uniformity, which are very important for large area prints required for cathode and anode application. For screen printing commercially available sintered electrolytes with dimensions of 50x50 mm^2 made of 10Sc1CeSZ were used. The thickness of the substrates varied from 110 µm to 210 µm. The electrochemical experiments were carried out on the cells with symmetrically screen-printed cathode and anode having lateral dimensions of 40x40 mm^2. A multilayer anode and a two-layer cathode were used.

The first cathode layer (~20µm) was composed of mixture lanthanum strontium manganite with or without additional transition metal on B-place LSMM´ and ScSZ. The second layer (~25 µm) consisted of LSMM´ and had a current collector function.

The first anode layer can be mainly composed of electrolyte material and helps for adhesion of upper anode layers to the substrate. The next anode layers are more electrochemically active and the upper anode layer with higher volume content of NiO is a current collector. The total anode thickness was <35 µm. The electrodes were sintered in co-firing at temperatures <1325°C.

The morphology of the studied electrodes was analyzed using field emission scanning electron microscopy.

Additional electrochemical tests were carried out with symmetrically screen-printed cathodes, which were also characterized by impedance spectroscopy at open-circuit potential (temperatures of 700-950°C) with a Schlumberger SI1260 impedance analyzer in dependence of the oxygen partial pressure. For this experiment the substrates with dimension of 15x15 mm^2 were printed symmetrically with cathode (with lateral dimension of 5x5 mm^2) and sintered in air at same conditions like complete cell. After sintering the symmetrical cells were painted with gold paste (Fraunhofer IKTS Dresden, Germany). The electrolyte resistance was taken from the high frequency end of the Nyquist plot (f=56 kHz). The contact resistance was proved to be <2 mΩ.

All electrochemical experiments were carried out using the advanced test bench for cell characterization developed at Fraunhofer IKTS, Dresden. The apparatus consists of a custom-built ceramic housing integrated in a furnace operated by a temperature controller enabling the temperature-time profile management up to 1100°C. Pt and Ni meshes were used as contact material for cathode and anode respectively. To provide the perfect contact between cathode and Pt mesh of current collector the cathode of the sintered cell is coated with contact paste by screen printing and the other Pt mesh is inserted into the screen printed paste.

The developed SOFC cells were characterized by impedance spectroscopy under current load (up to 650 mA/cm^2) at temperatures of 700-950°C in air:hydrogen/steam dual atmosphere using impedance analyzer IM6 (Zahner, Germany). The frequency was varied between 10 mHz and 100 kHz, the excitation AC voltage was fixed at 10 mV. The constituents of anode, cathode and electrolyte in the overall resistance are extracted from impedance spectra using Thales® Software (Zahner, Germany) and adequate equivalent circuit.

The influence of the different pH$_2$O concentration in anode gas, fuel and oxygen utilization on the impedance spectra and electrochemical activity of developed cells was investigated.

The current-voltage plots with current rate of 0,2A/min up to 20A and impedance spectra have been measured at different experimental conditions. The ASR of the cell has been corrected to fuel utilization and calculated using equations represented in [14].

RESULTS & DISCUSSION

Symmetrical cell

The typical impedance spectra of cells with LSM/10Sc1CeSZ cathode at T≤850°C are shown in the Figure 10. The capacitance values calculated from equivalent circuit in Fig. 9 shows a good agreement with previously estimated values for $C_{I.1}$, $C_{I.2}$ and $C_{I.4}$. We have seen that $C_{I.1}$ had a weak dependence on both temperature and oxygen partial pressure which is in a good agreement with equation (6). $C_{I.2}$ raised slightly decreasing the temperature (weak Arrhenius type dependence) for LSM electrode and was almost constant for LSMM' electrode. For both electrodes the increase of $C_{I.2}$ due to raise of L_c reducing oxygen partial pressure in agreement with Eq. 18 was observed. $C_{I.4}$ growed increasing temperature.

The similar behaviour was observed for LSMM'/10Sc1CeSZ electrodes with exept to $C_{I.2}$.

The increase of $C_{I.4}$ with temperature we explain by the fact that this capacitance is proportional to interfacial area TPBL*L_c and should increase with the extension of TPB (L_c) inside the LSM grains increasing the temperature. These observations suggest that the processes in the LSM electrode take place on the surface with only small extension of electrochemically active zone inside the material.

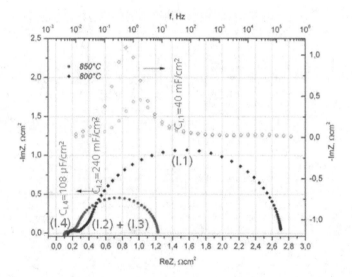

Fig. 10 Impedance spectra of LSM/10Sc1CeSZ cathodes at 800 and 850°C in air.

The comparison of impedance for LSM/10Sc1CeSZ and LSMM'/10Sc1CeSZ cathodes is shown in the Figure 12. The absolute values of polarization resistances for all steps of electrochemical reaction can be significantly reduced using perovskite composition with higher surface oxygen exchange and higher L_c. Due to optimization of paste preparation and sintering conditions also the high frequency impedance has been reduced for LSMM'/10Sc1CeSZ cathode by avoiding the zirconate formation at the perovskite/electrolyte interface. The area specific capacitances at 850°C for advanced electrode have much higher values ($C_{I.1}$=134 mF/cm², $C_{I.2}$=495 mF/cm² and $C_{I.4}$=840 µF/cm²) in comparison to standard LSM/10Sc1CeSZ composition ($C_{I.1}$=40 mF/cm², $C_{I.2}$=240 mF/cm² and $C_{I.4}$=108 µF/cm²). This change took place due to following reasons:

(i) $C_{I.4}$ increased due to increasing of interfacial area between e.c. and i.c. TPBL*L_c and by suppressing of ziconate formation (possibly thinner zirconate layer and double layer)

(ii) $C_{I.1}$ increased due to doping of LSM with transition metal catalyzing the oxygen surface exchange

(iii) $C_{I.2}$ increased due to increase of active electrode thickness (L_c) in Eq. 18 because of better catalytic activity of composite material

Not only material parameters are responsible for this gain in performance but also microstructure of the electrodes (the length of TPB). Figure 11 shows that for LSMM'/10Sc1CeSZ cathode the TPB length is substantially increased (we suppose at least by factor 1.3).

The PO_2 dependences of $R_{I.1}$, $R_{I.2}$ and $R_{I.4}$ obtained from impedance spectra using equivalent circuit in Fig. 9 show that $R_{I.4}$ doesn't depend on PO_2 for both composite cathodes (Figure 12). $R_{I.1}$ is strongly PO_2 dependent ($R_{I.1} \approx const \cdot P_{O_2}^{-0,6}$). The dependence is very close to the theoretically predicted relation between $R_{I.1}$ and PO_2 ($R_{I.1} \approx const \cdot P_{O_2}^{-0,5}$). We suppose that the deviation from theoretical behavior is caused by non-symmetrical reaction character for reaction I.1 (cathodic direction is more favorable than anodic one and therefore =0,6 in Eq. (9)).

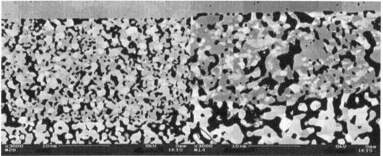

Figure 11. Microstructure of LSM/10Sc1CeSZ (right) and LSMM'/10Sc1CeSZ (left) cathode.

Surprisingly $R_{I.2}$ for LSM and LSMM' cathodes showed different dependence on PO_2. Whereas LSMM' cathode performed in accordance with theoretically predicted $R_{I.2} \approx const \cdot P_{O_2}^{-0,25}$ rule for surface driven kinetics there was no PO_2 dependence of $R_{I.2}$ for LSM cathode. We explain this behaviour of LSM based cathode by dependence of   (number of the adsorption sites per unit area) from oxygen vacancy concentration in LSM. This observation is in a good agreement with observed activation of LSM based cathodes under cathodic overpotential. This causes the weak dependence of $R_{I.2}$ from PO_2 according Eq. (13). It means that enhancing the oxygen vacancy concentration in LSM by cathodic overpotential the surface exchange can be highly activated and the polarization resistance lowered.

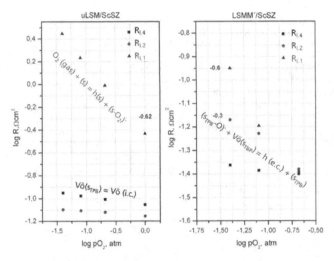

Figure 12. PO$_2$ dependence of polarization resistance for LSM and LSMM' based cathode at 850°C.

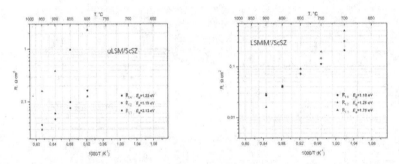

Figure 13. Temperature dependence of polarization resistance for LSM and LSMM' based cathodes in air.

The temperature dependence of all constitutents of cathodic polarization resistance is shown in Fig. 13. It can be clearly seen that especially the activation energy for oxygen exchange between air and catalyst surface as well as for oxygen transfer from TPB into ionic conductor have been reduced for LSMM'/10Sc1CeSZ composite. It is also clearly seen that $R_{I,1}$ has an activation energy typical for surface exchange and $R_{I,4}$ the typical activation energy for ionic conductivity. Because the activation energy for surface reactions for LSMM' is less than that for LSM (see Fig. 13) the $C_{I,2}$ dependence on temperature is for LSMM' weaker (see discussion to Eq. 18).

Impedance of MEA with Ni/YSZ and Ni/CGO anode

The impedance spectra of cells with Ni/YSZ and Ni/CGO anode are shown in Fig. 14. The deconvolution of spectra has been made using equivalent circuit in Fig. 8. In the MEA spectra we doesn't observe the high frequency response of cathode ($R_{I.4}$ and $C_{I.4}$). We suppose that after activation of the cell with current the zirconate is destroyed and the resistance $R_{I.4}$ reduces (disappears). It was found that the reaction on Ni/YSZ anode can be easily described using two additional arcs for reaction II.1 and II.2. Unfortunately the impedance arcs for processes I.1 and II.1 appears at the similar frequencies and can not be separated under chosen measurement conditions. For low frequency arc we obtain the capacitance in the range of 750 mF/cm² (which is a product of hydrogen and oxygen adsorption in the anode or cathode respectively) and in the high frequency arc we observe a capacitance of $C_{II.4}$= 200 µF/cm² for Ni/YSZ interface. $C_{II.4}$ increases increasing the operating temperature of the cell and low frequency arc capacitance seems to stay more or less constant.

In Ni/CGO electrode some differences to Ni/YSZ anode were observed:

(i)   the high frequency capacitance had higher values compared to Ni/YSZ anode (ca. 2 mF/cm²)

(ii)   low frequency capacitance is in the range of 1 F/cm² what is definitly higher than capacitance for reaction I.1 measured on symmetrical cells (ca. 500 mF/cm²)

(iii)   additional electrochemical process with capacitance of 160 mF/cm² at ca. 50 Hz is found. This capacitance was not dependent from temperature and H_2O content in the fuel (5-50%).

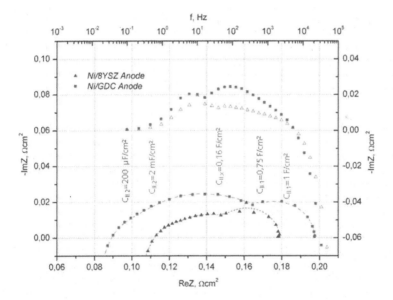

Figure 14. Impedance spectra of MEAs with LSMM'/10Sc1CeSZ cathode and Ni/YSZ and Ni/CGO anode at 850°C with air as oxidant.

Fig. 15 shows the performance of both anodes in dry reformate in comparison to operation in moistured hydrogen. The sensitivity of Ni/YSZ to H_2O content in the range of high frequencies can be clearly seen. Ni/CGO anode performance has not changed reducing the humidification. At low frequency the gas conversion impedance appears. The observed values of $R_{II,Nernst}$ and $C_{II,Nernst}$ are in a good agreement with values calculated in Table II.

No changes in impedance spectra from the current density using $H_2:H_2O=1:1$ as a fuel were observed.

Based on these observations we suppose that the electrochemical reactions in Ni/CGO anode are different from that in Ni/YSZ anode and take place mostly on the CGO surface. The capacity of the new electrochemical process is too high for grain boundary electronic or ionic (protons or oxygen ions) conductivity. We suppose that this capacity is associated with oxygen capacity of ceria similar to that of mixed ionic conductor in the cathode (see Eq. 14).

From the measurements at different temperatures we have found the activation energies for different constituents of MEA impedance. We have found that the cathodic reaction has been strongly activated by passing the current through the cell and the activation energy as well as absolute values of polarization resistances are smaller in comparison to that obtained on symmetrical cells. This effect can be explained by introduction of additional vacancies in LSM or LSMM' and destruction of zirconates at the interfaces during passing the current through the interface. We have also found that the composite cathode itself has much lower polarization resistance and activation energy than pure perovskite electrode due to enrichment of perovskite particles with additional oxygen vacancies during interaction with electrolyte by sintering. The high frequency anodic impedance arc has an activation energy in the range of that of the ionic conductivity. The new frequency arc has relatively low activation energy.

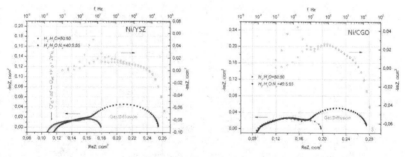

Figure 15. Impedance spectra of MEAs with LSMM'/10Sc1CeSZ cathode and Ni/YSZ and Ni/CGO anode at 850°C with air as oxidant in moistured and less humidified fuel.

We propose following reaction mechanism for Ni/CGO anode

$$H_{2(g)} + s_{CGO} \leftrightarrow 2H_{ad} \qquad\qquad \text{III.1}$$

$$H_{ad} + O^{2-}_{CGO} \leftrightarrow OH_{ad} + s_{CGO} + e^-_{CGO} \qquad\qquad \text{III.2}$$

$$H_{ad} + OH_{ad} \leftrightarrow H_2O_g + s_{CGO} + e^-_{CGO} \qquad\qquad \text{III.3}$$

$$O^{2-}_{CGO} \leftrightarrow O^{2-}_{YSZ} \qquad\qquad \text{III.4}$$

$$e^-_{CGO} \leftrightarrow e^-_{Ni} \qquad\qquad \text{III.5}$$

We consider reactions III.3 for low frequency process, reaction III.2 for new observed process at 50Hz and reaction III.4 or III.5 for high frequency process at CGO/YSZ or CGO/Ni interface respectively. This hypothesis should be proved further more carefully in our further research.

CONCLUSIONS

The electrochemical impedance spectra of LSM-based perovskite/10Sc1CeSZ cathode can be consistently explained using two rate determining steps: oxygen exchange on the surface of perovskite and oxygen reduction at TPB. It was shown that for LSM-based perovskites the surface kinetics is dominant for T≤850°C. The impedance spectra can be described by low (oxygen adsorption) and high (oxygen vacancy transfer from TPB into electrolyte) frequency RC elements and Gerisher impedance. Using this simple equivalent circuit it was possible to fit impedance spectra of symmetrical cells in temperature range 700-850°C. The capacitance absolute values as well as their dependence on PO_2 and temperature for every step of cathodic reaction could be good correlated to results of simplified models for single step processes.

The impedance spectra of MEA with Ni/YSZ anode was investigated in the P_{H2O} region where the conversion impedance could be neglected. The spectra is fitted introducing two additional RC-elements for dissociative hydrogen adsorption in temperature range 700-850°C giving consistent dependence of polarization resistance and capacitance from the temperature. The low frequency arc for hydrogen adsorption overlapped with the arc originated from oxygen adsorption on the cathode side and can not be clearly separated. The high frequency arc of anodic impedance could be well seen and showed dependence on hydrogen partial pressure. The interfacial reactions in LSM were strongly activated by passing the current through the MEA. This activation is caused by dissolution of zirconates (nanolayers) at the LSM/10Sc1CeSZ interface and generation of additional vacancies in perovskite.

Application of Ni/CGO anode changed the appearance of impedance spectra: high frequency process was shifted to lower frequencies and new process at ca. 50 Hz has been found. Probably the change of the mechanism of the electrochemical reaction takes place due to use of doped ceria with sufficient amount of electronic conductivity. The mechanism of hydrogen reduction on ceria which would explain the observed spectra has been proposed.

REFERENCES
[1] S.B. Adler, J.A. Lane, B.C.H. Steele, Journal of Electrochemical Society, 143 (1996) 3554
[2] A. Weber: Entwicklung von Kathodenstrukturen für die Hochtemperatur-Brennstoffzelle SOFC. Ph. D. Thesis, Universität Fridericiana Karlsruhe, 2002
[3] S.P. Jiang, S.H. Chan, Journal of Materials Science, 39 (2004) 4405
[4] T. Setoguchi, K. Okamoto, K. Eguchi, H. Arai, Journal of Electrochemical Society, 139 (1992) 2875
[5] A. Bieberle, Journal of the Electrochemical Society, 148 (6) (2001) A646
[6] A. Bieberle, The Electrochemistry of SOFC Anodes, Dissertation, ETH Zurich, (2000)
[7] H. S. Spacil. Electrical device including nickel-containing stabilized zirconia electrode. US Patent, Pages 3,558,360, (1970)
[8] A. Weber, D. Klotz, V. Sonn, E. Ivers-Tiffeé, Impedance Spectroscopy as a diagnosis tool for SOFC stacks and systems, International symposium on diagnostic tools for fuel cell technologies, Trondheim, Norway, (2009).
[9] B.C.H. Steele, Solid State Ionics 75 (1995) 157
[10] B.C.H. Steele, Solid State Ionics 86-88 (1996) 1223
[11] Anne C. Co, Shen Jiang Xia, and Viola I. Birss, Journal of the Electrochemical Society, 152 (3) (2005) A570
[12] M. Kuznecov, P. Otschik, P. Obenaus, K. Eichler, W. Schaffrath, Solid State Ionics 157 (2003) 371

[13] M. Kellner, T. Betz, M. Kusnezoff, N. Trofimenko and S. Mosch, in 8th European SOFC Forum, Lucerne, 30 June- 4 July 2008, B0501 (2008)

[14] A. Glauche, T. Betz, S. Mosch, N. Trofimenko and M. Kusnezoff, ESC Trans., 25 (2), 411 (2009)

[15] S. Mosch, N. Trofimenko, M. Kusnezoff, T. Betz and M. Kellner, in 7th European SOFC Forum, Lucerne, 3 - 7 July 2006, P413 (2006)

THE EFFECT OF A-SITE STOICHIOMETRY ON LSCF CATHODE PERFORMANCE AND STABILITY

Jared Templeton, John Hardy, Zigui Lu, Jeff Stevenson
Pacific Northwest National Laboratory
Richland, Washington, 99352 USA

ABSTRACT
Anode-supported solid oxide fuel cells (SOFCs) were prepared and tested. The cells were identical except for variations in the A-site stoichiometry of the LSCF cathode. A commercial supplier provided lanthanum strontium cobalt ferrite (LSCF) powders that were stoichiometric [$La_{0.6}Sr_{0.4}Co_{0.2}Fe_{0.8}O_{3-\delta}$], A-site deficient [$(La_{0.6}Sr_{0.4})_{1-x}Co_{0.2}Fe_{0.8}O_{3-\delta}$], and Sr-deficient [$La_{0.6}Sr_{0.4-x}Co_{0.2}Fe_{0.8}O_{3-\delta}$], in which the imposed deficiency (x) was 0.02, 0.05, and 0.1 in the non-stoichiometric cases. Multiple 1,000 hour tests were run using each LSCF cathode composition. The performance and stability of the cells will be discussed.

INTRODUCTION
SOFCs are an attractive power source due to their high efficiency, but are currently too expensive to market as commercially viable products. The 700-800℃ operating temperatures of intermediate temperature (IT-)SOFCs has enabled the use of stainless steels in the SOFC framework and current collectors, which results in significant cost reductions. However, the lower operating temperatures of IT-SOFC's also cause significant decreases in power densities of cells with LSM cathodes due to their high activation energies. LSCF is a mixed ionic-electronic conducting perovskite that exhibits higher performance than LSM/YSZ composites and shows potential as a replacement cathode [1].

EXPERIMENTAL
Ni-YSZ anode supported YSZ thin film bi-layers composed of bulk anode, active anode, and electrolyte were created via a tape casting and subsequent lamination process and has been described previously [2]. The resultant green tapes were cut into 29 mm circles and sintered at 1375℃ for 2 hours and then creep flatted at 1350℃ for 2 hours. A barrier layer consisting of $Sm_{0.2}Ce_{0.8}O_{1.9}$ (samarium-doped ceria, SDC, Praxair Specialty Ceramics) was added between the YSZ electrolyte and cathode to block chemical reactions and inter-diffusion between the two layers. The SDC barrier layer was screen printed onto the YSZ surface and sintered at 1200℃ for 2 hours. ($La_{0.6}Sr_{0.4})_{1-x}Co_{0.2}Fe_{0.8}O_{3-\delta}$ (x = 0, 0.02, 0.05, 0.10) and $La_{0.6}Sr_{0.4-y}Co_{0.2}Fe_{0.8}O_{3-\delta}$ (y = 0.02, 0.05, 0.10) powders from Fuel Cell Materials (FCM, Lewis Center, OH) were used as the cathode materials. Each as-received cathode powder was attrition milled to reduce the particle size diameter to between 0.35-0.40 microns. LSCF cathodes with a diameter of 2 cm^2 were screen printed onto the SDC barrier layer and sintered at 1100℃ for 2 hours. All cathode and SDC inks used in screen printing were made using Heraeus V-006 binder at 40% solids loading and mixed in a three roll mill. $La_{0.8}Sr_{0.2}CoO_3$ (LSC) that was made using the glycine-nitrate process [3] was screen printed onto the LSCF surface and embedded with Au gauze for cathode current collection. NiO grids with embedded Ni gauze were used as current collectors for the anode.

Each cell was sealed to an alumina tube test stand using a glass seal and tested at 750℃. Hydrogen with 3% H_2O was supplied to the anode at 200 sccm while air flowed to the cathode at 400 sccm. I-V and impedance spectra were recorded using a Solartron 1470 Multistat in conjunction with a Solartron 1255B frequency response analyzer. Cells were tested under two different constant current conditions. First, tests were carried out at the current that was initially generated by each cell at an operating voltage of 800mV. Then, additional cells were tested at a constant current of 1.937A, which was the average current generated at 800 mV at start up as calculated from a population of cells with the six different non-stoichiometric LSCF compositions being investigated in this study (3 cells with each composition for an 18 cell average). Each test was run for 900-1000 hours with impedance spectroscopy performed every 25-100 hours. After testing, the fuel cells were observed under scanning electron microscopy (SEM) (JEOLJSM-5900LV) for cross sectional viewing and energy dispersive x-ray spectroscopy (EDS) analysis.

RESULTS AND DISCUSSION

Methodology
        Electrolyte supported SOFC's provide the advantage of being able to use reference electrodes to separate the impedance losses incurred by the cathode and anode portions of the cell during operation. However, due to the high resistivity of the thick electrolyte support, these cells are unable to generate the high current densities generated by anode supported thin film electrolyte cells. Anode supported cells offer higher power densities, but do not allow the separation between anode and cathode impedance losses.
        If the SOFC substrates (bulk anode support, active anode, and electrolyte layers) on which the cathodes are deposited could be made to be identical and were tested at the same constant current and temperature in the same fuel gas environment (fuel composition and flow rates), it could be assumed that the anode's contribution to polarization for each of the cells would be the same. Although it is impossible to make the SOFC substrates for multiple cells absolutely identical, the differences between them can be minimized by making cells using substrates made from the same tape cast batches, tapes, and/or laminates. In this study, although the anode impedance cannot be quantified and eliminated from the total impedance of the cell, all of the conditions described above (nearly identical substrates, same fuel gas composition, same temperature and same constant current) are being employed to normalize it between the tested cells to the extent possible. If the assumptions hold true, any observed difference in total cell impedance can be attributed to the cathode layer and its interaction with the barrier and electrolyte layers. This method provides a way to compare the performance and stability of various cathode materials against each other by minimizing differences due to anode impedance contributions.
        The plots of power density as a function of time for the tested cells were fitted using SigmaPlot 2008 (v11.0, SPSS Inc.) software. A five parameter double exponential decay Equation (1) was found to give the best fit with R^2 values above 0.97. In Equation (1), p = power density, t = time in hours, and a, b, c, d, and p_0 are fitting parameters.

$$p = ae^{-bt} + ce^{-dt} + p_0 \qquad (1)$$

        This best fit equation allows for the calculation of instantaneous degradation rates of the cells as a function of time. The degradation rate is the instantaneous slope of the power density curve and is equivalent to the derivative of Equation (1) evaluated at time t. The derivative of Equation (1) is given in Equation (2). The degradation rate in %/1000 hours can then be expressed by Equation (3).

$$\frac{dp}{dt} = -abe^{-bt} - cde^{-dt} \qquad (2)$$

$$r = \frac{-1000\frac{dp}{dt}}{p(t)} * 100\% \qquad (3)$$

Power Density and Stability

Figure 1 shows the power density as a function of time for 4 series of tests on the 7 LSCF cathode compositions, two at 800mV and two at 1.937A constant current conditions. LSCF cathode compositions with varying A-Site and Sr-deficiencies operated at a significantly higher power density in both conditions than stoichiometric LSCF powders from both Praxair Specialty Ceramics (Woodinville, WA) and FCM. In addition, while performing better than stoichiometric LSCF powders, LSCF cathodes which were 10% Sr-deficient also performed much lower on average than other A-site and Sr-deficient compositions. The best fit curves are overlaid in black over the measured power density curves of each cathode composition. No consistent trend with cathode composition could be concluded from the results. With the possible exception of the 10% Sr-deficient LSCF, it appears as though the effects on cell performance due to compositional differences are masked by the experimental uncertainty. Although the effect of cathode composition on cell performance could not be clearly distinguished, it was obvious that A-site deficiency is beneficial for LSCF cathodes.

Figure 2 shows the degradation rates of the various cathodes when tested at 800 mV and 1.937A as calculated from curve fitting results using the method described above. Initial degradation rates are very high (>20%/1000 hours) for all compositions. After 900-1000 hours of testing, the degradation rates for stoichiometric, A-site and Sr-deficient cells were all between 2 and 6%/1000 hours. As with power density, there was no consistent trend between cathode composition and degradation rate.

Figure 3 shows back-scattered electron SEM images of polished cross-sections of a 2% A-site deficient and a 2% Sr-deffficient LSCF cathode. The cathode microstructures seen in Figure 3 are typical of the cathode compositions tested. Strontium is seen at the interface between the SDC and YSZ layers, where it is known to form insulating $SrZrO_3$. It can be assumed that part of the increase in ohmic resistance of the cells during operation is caused by the diffusion of strontium to the YSZ surface. Figure 4 is an elemental map showing Co distribution in a cross sectional area of 10% Sr deficient tested cathode. Regions of concentrated cobalt were found in the in the cathode while maps of other cathode elements appeared to be homogeneous. This Co segregation could account for the fact that the power density of 10% Sr deficient cells was lower on average than the power density of the other compositions.

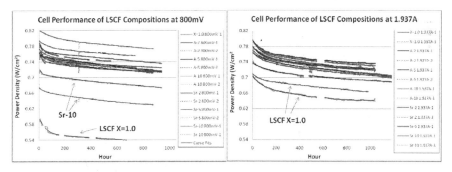

Figure 1. Power density vs. time of LSCF cathode compositions tested at 800mV (a) and 1.937A (b)

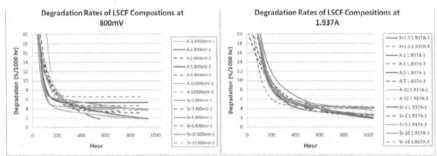

Figure 2. Degradation rates of LSCF cathode compositions tested at 800mV (a) and 1.937A (b)

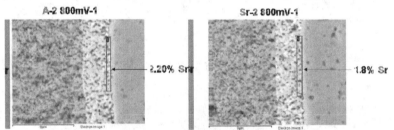

Figure 3. SEM image of 2% A-site (a) and 2% Sr-site (b) deficient cells tested at 750C at 800mV

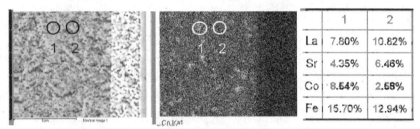

	1	2
La	7.80%	10.82%
Sr	4.35%	6.46%
Co	8.54%	2.58%
Fe	15.70%	12.94%

Figure 4. SEM image of 10% Sr deficient cell tested at 750C at 800mV

Electrochemical Impedance Spectroscopy

Impedance scans were performed every 25 to 100 hours on each of the cells during operation. Electrode polarization losses were calculated by subtracting the ohmic loss from the total cell resistances. Figures 5 and 6 show the ohmic and polarization loses of each LSCF composition tested at 800 mV(a) and 1.937A (b) respectively. It can be seen in general that the cell ohmic and polarization resistances are closely grouped with each other and there is no consistent trend with chemical composition.

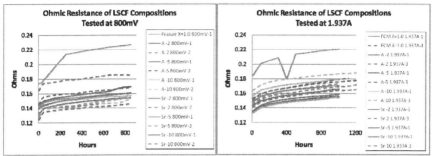

Figure 5. Ohmic Resistance of LSCF cathode compositions tested at 800mV (a) and 1.937A (b)

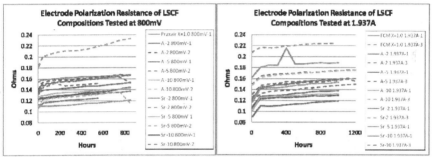

Figure 6. Electrode Polarization resistance of LSCF compositions at 800mV (a) and 1.937A (b)

CONCLUSION

Cell tests showed that A-site and Sr-deficient LSCF cathodes consistently outperform the power densities of stoichiometric LSCF cathode power densities by up to 10%. It is also shown that all stoichiometric, A-site, and Sr-deficient LSCF cathodes degrade over time at similar rates. Contributions of ohmic and electrode polarization losses to cell degradation rates were similar regardless of cathode composition.

ACKNOWLEDGMENTS

This study was funded by the Department of Energy's Solid-state Energy Conversion Alliance (SECA) Core Technology Program.

REFERENCES

[1]T. Striker, J.A. Ruud, Y. Gao, W.J. Heward, C. Steinbruchel, A-site deficiency, phase purity and crystal structure in lanthanum strontium ferrite powders, Solid State Ion., 178, 1326-1336 (2007).
[2]S.P. Simner, M.D. Anderson, L.R. Pederson, J.W. Stevenson, Performance Variability of La(Sr)FeO3 SOFC Cathode with Pt, Ag, and Au Current Collectors, J. Electrochem. Soc., 152, 1851-1859 (2005).
[3]L.A. Chick, L.R. Pederson, G.D. Maupin, J.L. Bates, L.E. Thomas, G.J. Exarhos, Glycine-nitrate combustion synthesis of oxide ceramic powders, Mater. Lett., 10, 6-12 (1990).

INFLUENCE OF OPERATIONAL PARAMETERS ON LSCF AND LSF STABILITY

Amaia Arregui[1,2], Lide M. Rodriguez-Martinez[2], Stefano Modena[3], Jan van Herle[4] Massimo Bertoldi[3] and Vincenzo M. Sglavo[1]
[1]DIMTI, University of Trento, via Mesiano, 77, 38123 Trento (TN), Italy
[2]IKERLAN S.Coop, Pº. J. Mª. Arizmendiarrieta, 2, 20500 Mondragń, Spain
[3]SOFCPOWER S.p.A., via Al Dos de la Roda, 60, 38057 Pergine Valsugana, (TN), Italy
[4]Swiss Federal Institute of Technology Lausanne, EPFL, STI-ISE-LENI CH-1015 Lausanne, Switzerland

ABSTRACT

$(La_{0.6}Sr_{0.4})_{0.95}FeO_{3-\delta}/ Sm_{0.2}Ce_{0.8}O_2$ (70:30 w/w) and $(La_{0.6}Sr_{0.4})_{0.995}Co_{0.2}Fe_{0.8}O_{3-\delta}/Gd_{0.1}Ce_{0.9}O_2$ (50:50 w/w) cathodes screen printed on anode-supported cells were tested for 500 h under different conditions by using a Taguchi matrix that combined cathode processing parameters (cathode material, thickness, sintering temperature) and operation conditions (temperature, current density, air flux, air humidification and chromium presence). Individual effect of each control factor on cell performance degradation was evaluated: chromium poisoning, air humidity and temperature were classified, in such order, as the most influencing parameters for cathode and cell degradation. Unexpected high degradation rates were estimated at 750℃. Limited influence of the type of cathode on long-term behavior of cells was found; LSF/SDC cathodes exhibited, however, higher long-term stability.

INTRODUCTION

In recent years, significant efforts have been focused in the field of solid oxide fuel cells to lower the operation temperatures. Operating at intermediate temperatures (600℃-800℃) offers the possibility to use cheaper ferritic stainless steels as interconnect material, simplifies heat distribution, limits thermal expansion mismatch and inhibits thermodynamically activated reactions within the stack. In this regard, LSCF [(La,Sr)(Co,Fe)O$_{3-\delta}$] and LSF [(La,Sr)FeO$_{3-\delta}$] cathodes constitute promising solution due to their mixed ionic-electronic conductive nature as well as enhanced electro-catalytic activity compared to conventional LSM[1-6]. Nevertheless, LSCF and LSF cathodes have revealed intrinsic degradation mostly due to the perovskite instability and Sr depletion[7]. As far as chromium poisoning is concerned, LSCF and LSF mixed conductors, exhibit higher tolerance than LSM/YSZ cathodes, as a result of extended electrochemical active area that is not restricted to TPB (Triple Phase Boundary)[7].

Recently, degradation rate of stacks have been reduced to values around 0.5%(with LSM) and 3%(with LSCF) in 1000 h tests, this pointing out the necessity of studying the degradation mechanisms of single cells in more detail[8]. It is worth saying that the amount and nature of the degradation highly varies with operation conditions; experts in the field, however, do not agree in establishing optimal operating parameters. In the present work, a detailed comparison of the state of the art cathode materials used in intermediary operation conditions, LSCF/GDC and LSF/SDC, is presented.

Long-term experiments performed by varying control factors through a specific design of experiment (DoE) approach offer great advantage with respect to the classical "one factor at a time" approach. In the field of SOFC and especially in long-term stability research studies, DoE methodology allows time optimization as well as evaluating interactions that are not considered in a single parameter variation.

EXPERIMENTAL PROCEDURE

Commercially available anode-supported planar half-cells produced by Sofcpower S.p.a. were used in the present study. Circular half-cells (diameter = 30 mm) are produced by co-casting and sintering Ni/YSZ cermet anode (thickness = 220±20 μm) and 8YSZ (8 wt% Y$_2$O$_3$ – stabilized zirconia) electrolyte (thickness 8±2 μm). Such half-cells, after cathode deposition and subsequent sintering have been tested for more than thousand hours in single cell tests and stack operations, demonstrating stable performance.

In the present work, gadolinia-doped ceria (GDC, $Gd_{0.1}Ce_{0.9}O_2$) was adopted as a barrier to prevent chemical reaction between cathode and electrolyte and resultant $SrZrO_3$ formation. GDC powder was screen printed on the electrolyte after dispersing the powder in a terpineol and ethyl cellulose solution and fired at 1350°C for 2 h. The ceria layer thickness after sintering was ≈4μm.

The long-term stability of two different composite cathodes (i) LSF/SDC and (ii) LSCF/GDC was investigated. LSF/SDC and LSCF/GDC cathode samples were prepared by dispersing the composite cathode powder $(La_{0.6}Sr_{0.4})_{0.95}FeO_{3-\delta}/$ $Sm_{0.2}Ce_{0.8}O_2$ - 70:30 w/w - or $(La_{0.6}Sr_{0.4})_{0.995}Co_{0.2}Fe_{0.8}O_{3-\delta}/Gd_{0.1}Ce_{0.9}O_2$ - 50:50 w/w- respectively in terpineol and ethyl cellulose and screen-printing it on the GDC barrier layer. The cathode area of both type of cell was 1.96 cm². Two nominal electrode thicknesses were used (30μm and 50μm). This yielded to slightly thicker electrodes in the case of LSF/SDC composite powder. Cathodes were sintered at 950°C, 1000°C and 1050°C for 2 h in air. Table I illustrates the prepared combinations for each type of cathode. A $La_{0.5}Sr_{0.5}CoO_3$ layer (≈15μm) was screen-printed on the top of the cathodes after sintering to facilitate current collection. In addition, gold and nickel mesh were applied as current-collector at the cathode and anode side, respectively. An alumina filter was placed beneath the gold/nickel mesh to accommodate the 250 g/cm² pressure load.

In the experiments the chromium poisoning effect at the cathode was studied. In order to provide chromium vapors to the cell, pieces (14 mm x 14 mm x 0.5 mm) of commercial Crofer22 APU steel were placed between the alumina filter and the mechanical support (refractory material) of the current collector. The surface of the Crofer22APU steel was not pre-treated and was insulated from current collection in order to evaluate Cr-VI species diffusion to the cathode in 500 h tests. The cell was then finally fixed to the edge of an alumina tube with AREMCO 516 seal. The experimental setup is schematically shown in Fig. 1.

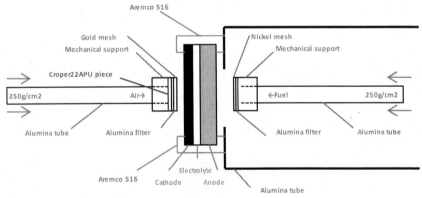

Figure 1. Experimental setup

Microstructure and morphology of the composite cathodes was examined by using scanning electron microscope (JEOL JSM 5500). Impedance spectroscopy measurements were acquired using Autolab PGSTAT302N high current potentiostat/galvanostat combined with the FR2A frequency response analyzer module. Impedances were measured in potentiostatic (0.7 V) conditions and frequency range was 10 kHz to 0.1 Hz with signal amplitude of 10 mV. The impedance data were normalized by the active contact area between the gold mesh and LSC current collector.

RESULTS

Table I reports obtained results in terms of *ASR* (area specific resistance), *Ro* (ohmic resistance), *Rp* (polarization resistance) and *P* (power density) as calculated from impedance spectroscopy measurements at 0.7V and 750℃, for cell combinations obtained from the Taguchi matrix.

Table I. Prepared cell combinations and impedance results at time 0 at 0.7 V, 750℃, 150 ml/cm ^{2} min air and 75 ml/cm^2 min H$_2$ (3% H$_2$O)

Cathode			Initial performances at 0,7V and 750°C			
Type	*Nominal Thickness(μm)*	*Sintering Temperature(°C)*	*ASR(Ωcm^2)*	*Ro(Ωcm^2)*	*Rp(Ωcm^2)*	*P(W/cm^2)*
LSF/SDC	30	950	0.48	0.31	0.17	0.311
		1000	0.40	0.20	0.21	0.294
	50	950	0.35	0.20	0.15	0.355
		1000	0.29	0.17	0.12	0.417
		1050	0.26	0.14	0.12	0.492
LSCF/GDC	30	1000	0.25	0.15	0.100	0.51
		1050	0.22	0.14	0.090	0.565
	50	950	0.24	0.17	0.080	0.507
		1000	0.22	0.13	0.090	0.581
		1050	0.16	0.11	0.05	0.722

Initial performance: influence of processing parameters

Impedance spectroscopy measurements led to lower ohmic as well as polarization resistances, for thicker cathodes as shown in Fig. 2. Despite of some scattering of obtained data, the general trend was unaffected for cathodes with similar composition and sintering temperature. Experimental results can be fitted with the model proposed by Liu by using equivalent circuits which describe the porous mixed ionic-electronic electrode behaviour[9]. In the cited work, it was proposed that an improvement of the MIEC/gas interface effectiveness relates to an increase on electrodes thickness, when ionic transport is sufficiently fast.

From the results in Fig. 2, it is observed that ohmic resistance decreases when sintering temperature is increased. Improved ohmic resistance at higher cathode sintering temperatures is attributed to a contact enhancement in the cathode/barrier interface. Conversely, no relation between sintering temperature and polarization resistance was found; the data shows large scattering for different combinations of cathode material and sintering temperature. Not all possible cell processing parameters have been used through the Taguchi matrix leading to additional difficulty in sintering temperature effect clarification. Nevertheless, in the range between 1000℃ and 1050℃ both LSF/SDC and LSCF/GDC cathode display a similar behavior and *Rp* is lower for sintering temperature of 1050℃.

Microstructural analysis of the cathodes in initial state revealed dissimilar behavior for LSF and LSCF cathodes with respect to the sintering temperature, as shown in Fig. 3. In the case of LSF/SDC cathode, finest particles are sintered already at 950℃ and particle size increases with sintering temperature up to 1050℃. Interconnectivity increases with temperature but during the first 100 h of operation considerable activation effect has been observed for cells sintered at 950℃. Conversely, for LSCF/GDC type cathodes, finer grains observed at 1050℃ are expected to deliver higher active surface where electrochemical activity occurs. In addition, the better connectivity within particles sintered at 1050℃ with respect to 1000℃, is in good agreement with the observed lower polarization resistance (Fig. 2). Experimental results reported here are in contradiction with conclusions made by Bebelis et al.[10], who attributed the improved performance of $La_{0.78}Sr_{0.2}Co_{0.2}Fe_{0.8}O_{3-\delta}$ with respect to $La_{0.58}Sr_{0.2}Co_{0.2}Fe_{0.8}O_{3-\delta}$ to the lowering of the sintering temperature from 1100℃ to 1060℃. However, in the mentioned work, the $La_{0.58}Sr_{0.2}Co_{0.2}Fe_{0.8}O_{3-\delta}$ cathode was ≈40μm thick whereas thickness of $La_{0.58}Sr_{0.2}Co_{0.2}Fe_{0.8}O_{3-\delta},$ was ≈30μm, as estimated from SEM images. This difference was not discussed in this work and can contribute to the cell performance.

At the same time, the superior performance for LSCF/GDC could be confirmed; both *Ro* and *Rp* values are lower for LSCF/GDC for cells with similar processing parameters. Thicker cathodes (≈50μm), exhibit higher power densities at 0.7V, as a consequence of both ohmic and polarization resistance decrease. Cathodes sintered at higher temperatures demonstrate better performances directly after sintering.

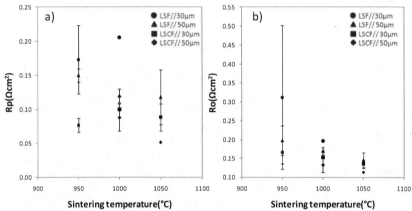

Fig. 2. Initial resistances from impedance spectroscopy measurements at 750℃, 150 ml/cm ^{2} min air and 75 ml/cm^2 min H$_2$; a) polarization b) ohmic resistances as a function of sintering temperature for different cathode composition and thickness

Taguchi matrix: control factors & levels

In the present study, a L18 Taguchi matrix which combines 8 control factors in 18 experiments was selected as shown in Table II. The original L18 orthogonal matrix consists of one control factor at two levels and seven control factors at three levels. Two of the control factors at three levels were downgraded to two levels through a Dummy treatment that preserves orthogonality of the matrix.

The considered control factors (processing parameters and operation conditions) that affect long-term stability of cell components and corresponding levels can be summarized as follows:

- Factor A: Cathode composition; Levels: A_1= LSF/SDC, A_2=LSCF/GDC;
- Factor B: Cathode thickness; Level: B_1=30, B_2=50μm;
- Factor C: Cathode sintering temperature; Level: C_1=950,C_2=1000, C_3=1050℃;
- Factor D: Presence of chromium source; Level: D_1= presence of Cr, D_2= no Cr;

- Factor E: Air flow; Level: E_1=30, E_2=75, E_3=150 ml/min cm^2;
- Factor F: Water vapor in cathode; Level: F_1=0, F_2=3, F_3=20% H_2O;
- Factor G: Current density; Level: G_1=0.25, G_2=0.5, G_3=0.75 A/cm^2;
- Factor H: Operation temperature; Level: H_1=700, H_2=750, H_3=800℃.

Fig. 3. SEM micrographs of LSF/SDC after sintering at (a) 950℃, (b) 1000℃ and (c) 1050℃ and of LSCF/GDC after sintering at (d) 1000℃ and (e) 1050℃

Table II. Taguchi design of experiments

	Processing parameters				Operation conditions			
Exp#	Cathode	t(μm)	T_s(°C)	Cr	Φ_{air}(ml/cm^{2}min)	%H_2O	J(A/cm^2)	T(°C)
	level	level	level	level	level	level	level	level
1	LSF-SDC	30	950	Yes	30	No	0.25	700
2	LSF-SDC	50	1000	Yes	75	3	0.5	750
3	LSF-SDC	50	1050	Yes	150	20	0.75	800
4	LSF-SDC	50	1050	No	30	No	0.5	750
5	LSF-SDC	30	950	No	75	3	0.75	800
6	LSF-SDC	50	1000	No	150	20	0.25	700
7	LSF-SDC	50	1050	No	30	3	0.25	800
8	LSF-SDC	50	950	No	75	20	0.5	700
9	LSF-SDC	30	1000	No	150	No	0.75	750
10	LSCF-GDC	50	950	Yes	30	20	0.75	750
11	LSCF-GDC	50	1000	Yes	75	No	0.25	800
12	LSCF-GDC	30	1050	Yes	150	3	0.5	700
13	LSCF-GDC	50	1000	No	30	3	0.75	700
14	LSCF-GDC	30	1050	No	75	20	0.25	750
15	LSCF-GDC	50	950	No	150	No	0.5	800
16	LSCF-GDC	30	1000	No	30	20	0.5	800
17	LSCF-GDC	50	1050	No	75	No	0.75	700
18	LSCF-GDC	50	950	No	150	3	0.25	750

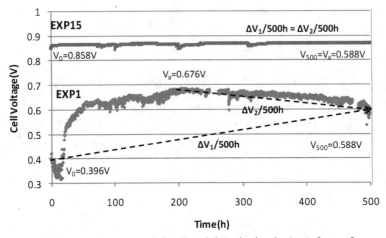

Figure 4. Degradation rate measurements using 0h and ah (activation time) as reference for experiment 1 and experiment 15

Long-term degradation analysis

Results from cell potential monitoring during long-term tests are shown in Table III for the Taguchi DoE in Table II. Cell potential versus time was characterized here by evaluating cell potential at zero, activation (ah) and 500 h points, as shown for experiment #1 and #15 in Fig. 4. Experiment #1 underwent appreciable activation during the first 200 h operation, followed by a period of constant voltage and, afterward, gradual performance loss occurred. Degradation rate measurements using zero point as reference do not distinguish however this degradation initiated at 200 h. In that sense, a second measurement from the maximal activation point is desirable to point out the degradation region that would otherwise be hidden in long-term tests of the actual characteristics. Degradation rates were calculated from experimental data by using both time zero and the maximal activation or the point where degradation starts, as the reference points. Degradation rates are summarized in the same Table II. The aim of this dual measurement is to distinguish the activation phenomena that occur in the first hundred hours of operation due to cathode sintering and contact improvement between the cathode and current collector. Both activation phenomena can strongly affect the tendencies in 500 h operation tests. All degradation rates are normalized to 500 h in order to guarantee fair comparison during data analysis.

The target function of interest for the optimization is the degradation rate at 500 h operation; $\Delta V/500h$. Individual results obtained from the Taguchi matrix combinations are plotted as a function of the corresponding number of experiment from table II in Fig. 5. The two minimum regions for experiment #1,2,3 and #10,11,12 correspond to LSF and LSCF electrodes, respectively, in contact with a chromium source. Also experiment #6 and #16 show considerable degradation rates around 200 mV in 500 h that are presumably due to air humidification. Degradation rate increases considerably in the case of experiment #1 and #10 when it is evaluated from the activation point instead of zero time.

Effect of the operation parameters on degradation

The experimental results from Fig.5 are analyzed by calculating the average degradation rate for each level of the control factors and are represented in the effect diagram in Fig. 6. The maximum difference between the factor levels represents how important the factor is and it is useful to identify the relative importance of the factorial effects, as shown in Fig. 7.

Table III. Potential monitoring results for the Taguchi design of experiments

Exp#	Operation data						Degradation rates	
	$t_0(h)$	$V_0(V)$	$t_a(h)$	$V_a(V)$	$t_{500}(h)$	$V_{500}(V)$	$\Delta V_1/500h$ (mV)	$\Delta V_2/500h$ (mV)
1	0	0.396	200	0.676	500	0.588	192	-147
2	0	0.797	12	0.822	499	0.480	-317	-350
3	0	0.757	5	0.770	190	0.626	-345	-388
4	0	0.802	83	0.816	478	0.795	-7	-27
5	0	0.696	90	0.710	500	0.551	-146	-194
6	0	0.917	30	0.926	500	0.917	0	-10
7	0	0.924	53	0.930	500	0.922	-3	-10
8	0	0.530	200	0.539	500	0.525	-5	-23
9	0	0.672	48	0.672	481	0.580	-96	-107
10	0	0.660	280	0.660	504	0.228	-429	-966
11	0	0.913	50	0.926	500	0.712	-201	-238
12	0	0.678	50	0.729	500	0.515	-163	-237
13	175	0.491	400	0.501	400	0.501	22	0
14	0	0.945	0	0.945	504	0.904	-40	-40
15	0	0.858	500	0.869	500	0.869	10	0
16	0	0.867	84	0.900	378	0.775	-121	-214
17	0	0.627	500	0.672	500	0.672	45	0
18	0	0.949	0	0.949	478	0.938	-12	-12

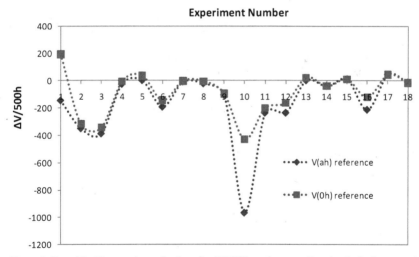

Figure 5. Taguchi table experimental values for $\Delta V/500h$ as the target function for both zero and activation time as reference.

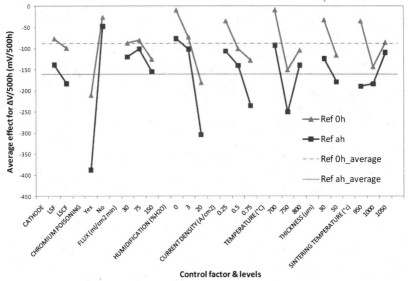

Control factor & levels

Figure 6. Individual factor effect for ΔV/500 h as the target function for both 0h and ah as reference

The presence of chromium vapors is identified as the principal factor that triggers degradation of cell performance (0h: 22% ; ah: %31), as seen in Fig. 7.

The second factor in the ranking is air humidification (0h: 20% ; ah: %21) and an average degradation rate as high as -300mV/500h was evaluated from the maximum activation point for 20 mol% humidified air, as observed in Fig.6. This results is in contradiction with experimental data reported by Nielsen et. Al.[11] who reported humidification tolerance for LSCF/GDC cathode humidified at 12.8 mol% during 100 h. Two differences can be however pointed out between Nielsen experiments and the configuration used in the present work; LSCF current collector was used in Risoe in place of LSC current collector and %0.5 A-site deficient LSCF cathode was used in the present work. Further analysis is required to determine air humidification influence for LSCF based cathodes.

The third important parameter is the operation temperature (0h: 17% ; ah: %14). The maximum of degradation at 750℃ was observed as shown in Fig. 6. This is probably the consequence of the overlay of two phenomena that show inverse trend with temperature; electrode polarization and chromium poisoning. The polarization resistance increases when temperature is lowered. However, chromium vaporization from the alloy is favored at higher temperatures. Such simultaneous degradation mechanisms could cause the observed maximum of degradation at 750℃. However, the magnitude of the degradation rates estimated at 750℃ remains surprisingly high.

Air flow rate had little influence on the degradation rates (0h: 5% ; ah: %5). The minimum degradation for 75ml/cm² min was found.

With regard to the current density, the optimal operation condition is 0.25 A/cm², the degradation tendency is rather linear and increases with the current load. These results are in good agreement with previous results[12-14].

Cells sintered at 950℃ are optimal if time 0 is taken as reference for degradation calculations but 1050℃ becomes the most suitable sintering temperature from the activation point suggesting an enhanced stability in prolonged tests.

Influence of cell thickness, air flux and cathode composition on degradation is not substantial but stability should be improved when 30μm thick cells in combination with 75ml/cm^2 min air through an LSF/SDC cathode are used.

Degradation mechanisms influenced by operation conditions as well as processing parameters effect can be further clarified by impedance spectroscopy measurements prior to and after the operation under the same conditions. Such results together with detailed post-mortem analysis are under investigation and expect to contribute in understanding the degradation mechanisms of LSF and LSCF cathodes.

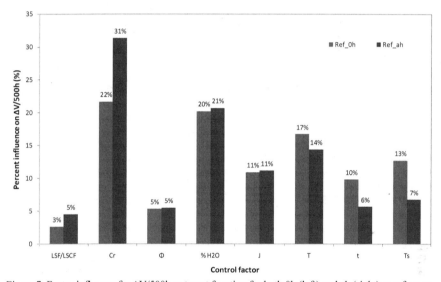

Figure 7. Factor influence for ΔV/500h as target function for both 0h (left) and ah (right) as reference.

CONCLUSIONS

The initial performances were better for cathodes with perovskites containing cobalt on the B-site. Higher power densities were collected for thicker cathodes (≈50μm); lower ohmic and polarization resistances were found for thicker cathodes with similar cathode composition and sintering temperature. On the other hand, for cathodes sintered at higher temperatures, 1050℃, better performance was observed; Ro decreased with increasing sintering temperature.

Operational parameters influence on long-term stability was successfully evaluated through a Taguchi matrix in 500 h tests. Chromium poisoning and cathode humidification were classified, in such order, as first and second most important parameter for cathode performance degradation. Temperature influence on degradation was also considerable and highest degradation rates were observed at 750℃. Further stability improvement was evaluated for low current loads, higher sintering temperatures and average air fluxes through the cathode. Modest influence of cathode composition was observed but LSF/SDC cathodes exhibited superior stability. Optimal parameters for minimum degradation in the investigated range are classified in Table IV.

Table IV. Optimal operation conditions for long-term stability

Processing parameters				Operation conditions			
Cathode	t(μm)	T_s(°C)	Cr	Φ_{air}(ml/cm^{2}min)	%H_2O	J(A/cm^2)	T(°C)
LSF/SDC	30	1050	No	75	0	0.25	700

REFERENCES

[1]R. Steinberger-Wilckens, I.C. Vinke, L. Blum, L.G. J. de Haart, J. Remmel, F. Tientz, W.J. Quadakkers, in *6th European Solid Oxide Fuel Cell Forum*, M. Mogensen, Editor, 1, pp. 11-19, European Fuel Cell Forum, Oberrohrdorf, Switzerland (2004).

[2]L.-W. Tai, M.M. Nasrallah, H.U. Anderson, D.M. Sparlin, S.R. Sehlin, *Solid State Ionics* 76 (1995) 259-271.

[3]L.-W. Tai, M.M. Nasrallah, H.U. Anderson, D.M. Sparlin, S.R. Sehlin, *Solid State Ionics* 76 (1995) 273-283.

[4]Y. Teraoka, H.M. Zhang, K. Okamoto, N. Yamazoe, *Mat. Res. Bull* 23 (1998).

[5]Y. Teraoka, T. Nobunaga, K. Okamoto, N. Miura and N. Yamazoe, *Solid State Ionics* 48 (1991) 207.

[6]S. Sekido, H. Tachibana, Y. Yamamura and T. Kambara, *Solid State Ionics* 37 (1990) 253.

[7]S.P. Simner, M.D. Anderson, M.H. Engelhard, J.W. Stevenson, *Electrochem. Solid-State Lett.* 9 (2006) A478–A481.

[8]H. Yokokawa, H. Tu, B. Iwanschitz, A. Mai, *Journal of Power Sources* 182 (2008) 400-412.

[9]M. Liu, *Journal of the Electrochemical Society*, Vol. 145, No. 1, (1998).

[10]S. Bebelis, N. Kotsionopoulos, A. Mai, F. Tietz, *Journal of Applied Electrochemistry*, 37 (2007) 15–20.

[11]L.-W. Tai, M.M. Nasrallah, H.U. Anderson, D.M. Sparlin, S.R. Sehlin, *Solid State Ionics* 76 (1995) 259-271.

[12]H. Yokokawa, H. Tu, B. Iwanschitz, A. Mai, *Journal of Power Sources* 182 (2008) 400-412.

[13]M.J. Heneka, E. Ivers-Tiffé e, in: S. Linderoth, A.Smith, N. Bonanos, A. Hagen, L. Mikkelsen, K. Kammer, D. Lybye, P.V. Hendriksen, F.W. Poulsen, M. Mogensen, W.G. Wang (Eds.), Proceedings of the 26th Risoe International Symposium on Materials Science, Riso National Laboratory, Roskilde, Denmark, 2005, pp. 215-222.

[14]H. Timmermann, Presentation on the International Workshop on Degradation Issues in Fuel Cell, Heronissos, Greece, September 19-21, 2007.

ASSESSMENT OF THE ELECTROCHEMICAL PROPERTIES OF BSCF AND SAMARIUM DOPED BSCF PEROVSKITES

Keling Zhang, Alex Lassman, Atul Verma[*] and Prabhakar Singh
Center for Clean Energy Engineering
University of Connecticut, Storrs, CT, USA

ABSTRACT

$Ba_{0.5}Sr_{0.5}Co_{0.8}Fe_{0.2}O_{3-x}$ (BSCF) and $(Ba_{0.5}Sr_{0.5})_{0.9}Sm_{0.1}Co_{0.8}Fe_{0.2}O_{3-x}$ (BSSCF) perovskites have been synthesized and characterized for their electrical conductivity and thermodynamic behavior as a function of temperature and oxygen activity. The electrical conductivity as a function of temperature in air for both compositions shows a maxima around 420 –450 ℃. The thermodynamic properties, the partial molar enthalpy, $\Delta \bar{H}_{O_2}$, and partial molar entropy, $\Delta \bar{S}_{O_2}$, were measured directly using an electrochemical cell for a fixed oxygen stoichiometry for both systems. The measurements show that BSCF undergoes a subtle polymorphic phase changes around 610 ℃ from cubic to hexagonal during cool down. It was found that Sm doping improves BSCF electrical conductivity and also stabilized the structure over the entire temperature range. The measured partial molar thermodynamic quantities were rationalized in terms of the unit cell volumes and phase stability.

INTRODUCTION

Acceptor-doped cobaltite perovskite-type oxides have been characterized by extensive lattice oxygen vacancy formation, which results in substantial departures from oxygen stoichiometry at higher temperatures[1-3]. A change in the vacancy concentration is accompanied by a corresponding change in the average valence of the B-site ions. The high concentration of oxygen vacancies in combined with their relatively high mobility brings about high oxygen ion conductivity. The intrinsic electronic conductivity is also high in these compounds which often become metallic at high temperatures. Due to the resultant high mixed conductivity, these materials also exhibit high ambipolar conductivity and oxygen permeability[4-6]. The combinations of these properties have been exploited for oxygen separation ceramic membranes[6] and cathodes for solid oxide fuel cells[2,3,7]. The promising candidates have been based on mixed oxygen ionic and electronic conducting oxides, such as $Sm_{0.5}Sr_{0.5}CoO_{3-x}$[8] and $La_{0.6}Sr_{0.4}Co_{0.2}Fe_{0.8}O_{3-x}$[9]. Recently, the perovskite $Ba_{0.5}Sr_{0.5}Co_{0.8}Fe_{0.2}O_{3-x}$ has attracted widespread attention for application as a cathode material in intermediate temperature solid oxide fuel cells (IT-SOFCs) and oxygen separation membranes[10-12].

Regarding optimization of the SOFC cathode and oxygen permeation membrane materials, there is a general agreement that a fast oxygen exchange reaction between the gaseous oxygen and the lattice oxygen, as well as high mixed conductivity in the bulk of material, is advantageous. The mixed conductivity extends the active oxygen reduction site from the typical TPB to the entire exposed cathode surface, thus greatly reducing the cathode polarization at low operating temperatures[13-15]. In the development of these materials, perovskites with transition metals at the B-site and partial or full substitution of rare earth metals by alkaline earth metals at the A-site have attracted considerable attention mainly due to their fast oxygen transport[16]. Among those compositions, $(La,Sr)(Co,Fe)O_{3-x}$ has been intensively investigated. While in this material solid solution series, the end member $SrCo_{0.8}Fe_{0.2}O_{3-x}$ (SCF) exhibited the highest oxygen permeation rate, the phase transition into the vacancy-ordered Brownmillerite phase at 750 ℃ in 0.015 atm was found detrimental for the application. To stabilize the cubic perovskite phase, Ba was introduced into $SrCo_{1-y}Fe_yO_{3-x}$ leading to $(Ba,Sr)(Co,Fe)O_3$. The cubic perovskite structure BSCF5582 in which 50% of Sr was substituted by Ba retains the cubic structure even under low oxygen partial pressure and temperature, and maintains high oxygen permeability at the same time[16,17]. Although the partial substitution of Sr^{2+} with Ba^{2+} in SCF led to an increase in the phase stability of the cubic perovskite structure and high oxygen ion conductivity, a decrease of the electronic conductivity was observed. The oxygen permeation study of the BSCF membrane demonstrated that the permeation rate was mainly determined by the slow surface exchange kinetics at the oxygen lean side of the membrane surface[18]. Besides the application as oxygen permeation membranes, BSCF also showed its applicability in partial oxidation of methane to syngas and ammonia oxidation[19]. The excellent mixed conducting properties and application in cathodes and ion transport membranes stems from both the thermodynamic and kinetic properties of the materials system. The materials system can tolerate large concentrations of mobile oxygen ions and the measured oxygen ion diffusivity and surface exchange co-efficient are much higher than those of comparable IT-SOFC cathode materials, such as SSC and LSCF[20]. Estimates based on the magnitude of the oxygen nonstoichiometry and the chemical diffusion coefficients indicate high ionic conductivities of 0.01–0.02 S cm^{-1} at 600–700 $\mathbb{C}$ [21,22]. These features support the application of BSCF in IT-SOFCs as cathodes with low polarization resistance even with reducing the operating temperatures towards 600 $\mathbb{C}$ [22]. BSCF was first assessed as SOFC cathode material with doped ceria as the electrolyte by Shao et. al which exhibited a high power density of 1010 mW/cm2 at 600 $\mathbb{C}$ with humidified hydrogen as the fuel[10]. The results have generated much interest and the material system has been studied by several groups in many other forms such as a composite and ion impregnated nanostructured cathodes[11]. The properties and performance of BSCF5582 and $Sm_{0.2}Ce_{0.8}O_{1.9}$(SDC) (70:30 in weight ratio) composite cathode for intermediate-temperature solid-oxide fuel cells have been were investigated. The observed improved performance was attributed to the enlarged cathode surface area contributed from the fine SDC particles[11].

There might be some intrinsic shortcomings of this material that prevent it from being wide spread use of this material, such as the formation of carbonate in the presence of CO_2, poor thermomechanical stability and a high temperature phase change from desired cubic phase to hexagonal phase[21]. The degradation in mixed conducting properties associated with this phase change has been observed by a number of research groups[23-25]. Furthermore, the electrical conductivity of BSCF is lower than the commonly used cathode such as $La_{0.8}Sr_{0.2}MnO3$, nearly one order of magnitude, which limits the conductance in the films composed of the material and also influences the electrochemical performance. A number of attempts has been made to modify the base composition of the BSCF material including A-site doping with Sm[26,27] and Pr[28] to improve the electrical conductivity. In this work, Sm doped BSCF, $(Ba_{0.5}Sr_{0.5})_{0.9}Sm_{0.1}Co_{0.8}Fe_{0.2}O_{3-x}$, has been synthesized and characterized.

The change in the thermodynamic state of BSCF is reflected by the oxygen released or uptake and a resulting deviation from stoichiometry, x. The macroscopic representation of the process is shown below by the solid-gas reaction:

$$Ba_{0.5}Sr_{0.5}Co_{0.8}Fe_{0.2}O_{3-x}(s) = Ba_{0.5}Sr_{0.5}Co_{0.8}Fe_{0.2}O_{3-x-\partial}(s) + \frac{\partial}{2}O_2(g) \qquad (1)$$

where ∂ is the deviation from the thermodynamic state resulting from the formation of point defects. The approach which has been used here is the defect equilibria for the formation and annihilation of ionic and electronic defects by the reaction of the oxide with the gas phase using a traditional formalism in which the relationships among defect concentrations and environmental parameters can be expressed by simple mass action type of equations. The latter are based on the assumption that all involved point defects can be treated as non-interacting and ideally diluted species for which corresponding chemical potentials can be defined. Such formalism has been applied successfully to model experimental data of oxygen nonstoichiometry and electrical conductivity of LSF[4], LSFeTi[29], LSFeCu[30] perovskites and has been extended here in this study with BSCF and BSSCF. Using an electrochemical cell constant composition technique, the partial molar thermodynamic quantities such as $\Delta\bar{H}_{O_2}$ and $\Delta\bar{S}_{O_2}$ can be directly obtained. Such measurements have been performed for pure and doped CeO_2[31]. Using the Gibbs energy formulation for the partial molar free energy

$$\Delta \bar{G}_{O_2} = \Delta \bar{H}_{O_2} - T\Delta \bar{S}_{O_2} \tag{2}$$

where the well known thermodynamic relations for constant composition can be used:

$$\Delta \bar{H}_{O_2} = \left(\frac{\partial \left(\frac{\Delta \bar{G}_{O_2}}{T} \right)}{\partial \left(\frac{1}{T} \right)} \right)_{n_j} \tag{3}$$

and,

$$\Delta \bar{S}_{O_2} = \left(\frac{-\partial \Delta \bar{G}_{O_2}}{\partial T} \right)_{n_j} \tag{4}$$

The above equation can be written in terms of cell emf, E, as $\Delta \bar{G}_{O_2} = -4FE$, and

$$\frac{\Delta \bar{H}_{O_2}}{RT} - \frac{\Delta \bar{S}_{O_2}}{R} = \frac{-4FE}{RT} = \ln P_{O_2} \tag{5}$$

The thermodynamic quantities, $\Delta \bar{H}_{O_2}$ and $\Delta \bar{S}_{O_2}$ can be easily deduced from plot of cell emf or Log P_{O2} vs. 1/T at constant composition. High temperature gas tight electrochemical cells employing solid electrolytes have been often applied to evaluate the thermodynamic properties. The constant composition technique offers a simple means for direct determination of partial molar quantities without any assumed defect model for the materials system which has been used to study BSCF and BSSCF. The high oxygen diffusivity and surface exchange rates of BSCF and BSSCF oxide materials have been reported, but chemical and structural stabilities have not been understood because of lack of reliable fundamental data. The phase change in BSCF and associated thermodynamic quantities the cubic perovskite phase which transforms to a hexagonal polymorph at lower temperatures has been studied. It was found that Sm doping improves BSCF electrical conductivity and also stabilized the structure over the entire temperature range.

EXPERIMENTAL
Material Synthesis: $Ba_{0.5}Sr_{0.5}Co_{0.8}Fe_{0.2}O_{3-x}$ (BSCF) and $(Ba_{0.5}Sr_{0.5})_{0.9}Sm_{0.1}Co_{0.8}Fe_{0.2}O_{3-x}$ (BSSCF) were synthesized using the EDTA-citrate process. First, the cation content of the metal-nitrate salts was measured, in order to ensure accurate synthesis. Known quantities of $Co(NO_3)_2 \cdot 6H_2O$ and $Fe(NO_3)_3 \cdot 9H_2O$ were heated to 800 ℃ for 5h in order to decompose the metal-nitrates into metal oxides. The resultant oxides were weighed, and the oxide formation was confirmed using XRD. Due to the reaction between Ba and the Al_2O_3 crucible during the heating process, a chemical titration technique was instead used to characterize $Ba(NO_3)_2$ and $Sr(NO_3)_2$, using EDTA as a complexing agent and Eriochrome Black T as the indicator.
After characterization, the metal-nitrates were dissolved in de-ionized water in the appropriate ratios. EDTA was added in a 1:1 metal:EDTA ratio, and citrate was added in a 1:2 metal:citrate ratio. Ammonium hydroxide was used to adjust the pH of the solution to 9. The water was evaporated by heating the solution to 60℃ on a stirring hotplate for three days, and the solution was then placed into an oven at 180℃ for 12h, after which it became a coarse powder. This powder was calcined at 1000℃ for 2h to form single phase BSCF powder. BSSCF powder required two additional grinding and recalcination steps to obtain single phase powder. X-ray diffraction of both the calcined BSCF and BSSCF powder and the sintered pellets were carried out on a Bruker D8 ADVANCE automated diffractometer, over a 2θ range of 20-80° with scanning rate 2°/minute, using Cu K α radiation, with generator voltage and current settings of 40 kV and 40 mA. Scanning electron microscopy (SEM)

analysis was carried out for both powders and sintered pellets with a FEI 250 Quanta scanning electron microscope.

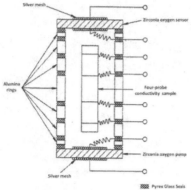

Figure 1. Schematic diagram of electrochemical titration cell with conductivity specimen

Ceramic pellets were fabricated by pressing powder into an approximate die pressure of 3 MPa using the Desk-top Electromotion Press (MTI CORPORATION). The pellets were then placed in high purity alumina boats and completely covered with additional powder for protection from contamination during sintering. A special firing schedule for sintering pallets was used which has been devised based on the rate of densification. The profile included slow heating around 850 ℃ an d extended dwelling times at 1050 and 1150 ℃. Crack -free dense samples were obtained after dwelling finally for 4 h at 1130 ℃. The pellets have a radius of 15 mm and thickness of 0.91mm. In order to form conductivity specimens, bars were cut from the pellet using SYJ-160 Low Speed Diamond Saw (MTI CORPORATION). BSCF bar has a dimension of 1.71 mm ×4.34 m m ×0.91mm, and BSSCF bar has a dimension of 3.33 mm ×3.60 m m ×0.91m m. Four shallow grooves were cut on the edges of each conductivity specimen. 5-mil silver wires were wrapped around each groove and served as electrodes. Silver conductive adhesive paste (Johnson Matthey, MA, US) was used to improve the contact between the sample and the wires, thus reducing the contact resistance. The two outer leads were used to flow a steady dc current while the inner electrodes served as the voltage probes. The painted samples were cured in-situ heated at 800 °C before the measurements. The measurements were performed in air upon heating from 300 to 800 ℃ using s teps of 25 °C. At each temperature step, 30 min was allowed for the temperature stabilization and equilibration in the gaseous atmosphere.
The sample conductivity, σ, was calculated as:

$$\sigma\left(\frac{S}{cm}\right) = \frac{I}{V}\frac{l(cm)}{A(cm^2)} \qquad (6)$$

where I is current, V is voltage drop, l is the inner electrode spacing, and A (ab) i s the cross-sectional area of the bar. Electrical conductivity was measured using the four-point technique in flowing air using Solartron 1287 potentiostat by current passing through the outer electrodes and measuring the voltage drop across the inner electrodes. The conductivity measurement in the electrochemical cell was also carried out by four-point technique using KEITHLEY 2400 Sourcemeter. The schematic diagram of this sealed cell chamber is given in Figure 1. The upper zirconia electrolyte as shown was used as an oxygen sensor and the lower zirconia electrolyte as an oxygen pump which is being used to change the partial pressure of O_2 inside the chamber. Sealing of the cell was accomplished in situ when the Pyrex

(Corning 7058) rings softened and squeezed out between the alumina rings during initial heat-up at 820 ℃. The sealing was tested by flowing two gases with different partial pressures of oxygen using the Nernest equation:

$$E = -\frac{4F}{RT}\ln\left(\frac{P_{O_2}}{P_{O_2\,ref}}\right) \tag{7}$$

In the equation, E is the electromotive force over the YSZ electrolyte disk, F is Faraday constant, R is the universal gas constant, T is the absolute temperature, P_{O_2} is the oxygen partial pressure inside the chamber, and $P_{O_2\,ref}$ is the oxygen partial pressure of the gas being flowed outside. A stable inside P_{O2} indicated a gas tight cell which was used in the thermodynamic evaluation of the perovskite materials. Two separate electrochemical cells were used for the two different compositions of the perovskites used in this study.

RESULTS AND DISCUSSIONS
 The phase purity of the starting powder materials were first examined using XRD patterns. Figure 2 below shows XRD patterns of pure BSCF and Sm doped BSCF powder and as seen the pattern matches well with the cubic perovskite peaks. From the results of XRD, it is evident clear that the substitution of the smaller Sm^{3+} (0.124 nm) substituted the larger Ba^{2+} (0.175 nm) and Sr^{2+} (0.158 nm) of the A-site cation causes the lattice shrinkage. The lattice shrinkage due to samarium doping the BSCF lattice has also been reported by Su et. al[26].
 Figure 3 shows the measured electrical conductivities as a function of temperature over a range of 300800 ℃. Electrical conductivity measurement revealed higher conductivity of BSSCF than BSCF.
The overall trend shows a maximum typical of perovskites and the low temperature region has been characterized by thermally activated p-type conductivity below the peak temperature. The high temperature region is the metallic conduction region where the conductivity decreases with temperature[32]. There has been disagreement between research groups regarding the magnitude conductivity and the transition temperature for BSCF materials in recent years.

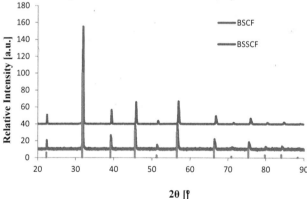

Figure 2. BSCF & BSSCF powder X-ray diffraction pattern. The peaks match cubic Barium Strontium Cobalt Iron Oxide - $Ba_{0.5}Sr_{0.5}Co_{0.8}Fe_{0.2}O_{3-x}$ (JCPD File No. 075-6980), shown in the green color.

Jiang et al.[33] observed a maximum of conductivity to occur around 470 ℃ both on heating and cooling for the BSCF material. Wei et al.[34] reported the electrical conductivity of BSCF as a function of temperature in both air and nitrogen atmospheres and observed that the material behaves differently above 450 ℃ depending on the ambient atmosphere. As the measured total conductivity of BSCF and BSSCF involves both ionic and electronic conductivity, due to the presence of oxygen vacancies and holes, it has strong dependence on the oxygen stoichiometry of the sample. It is likely that the difference in the magnitude and the peak conductivity is due to the differences in the sample stoichiometry, thermal history and concentration of triple and tetravalent state of Co and Fe cations. The BSCF and BSSCF compositions used in this study were equilibrated in air at 800 ℃ for 30 min and the conductivity data was collected during the cool down cycle at 25 ℃ intervals with 30 min equilibration time. The conductivity data show thermally activated p-type semi-conductivity below 420 and 450 ℃ with the highest value up to 57 and 75 S/m for BSCF and BSSCF, respectively. In Figure 3, conductivity is plotted as a function of temperature according to Equation 8, which is appropriate for conduction via adiabatic polaron hopping[6]:

$$\sigma = \frac{A}{T} e^{\frac{-Ea}{kT}} \qquad (8)$$

where Ea is activation energy, T is temperature in Kelvin, k is Boltzmann's constant, and A is the pre-exponential factor that depends on the site fraction of carriers and the probability that neighboring ions are available to participate in the hopping process.

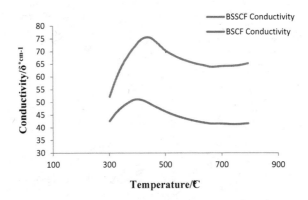

Figure 3. Total conductivity of BSCF & BSSCF as a function of temperature in air.

Below transition temperatures, the conductivity of BSCF and BSSCF is thermally activated with activation energies of 0.12 eV and 0.17 eV, respectively, which are comparable to those reported by others (Figure 4)[32,35]. The formation of electronic carriers beyond the peak temperature will alter the valences of the B-site cations and the formation of holes will convert Co^{3+} to Co^{4+} whereas the formation of electrons will convert Fe^{4+} to Fe^{3+}. Although both conversions are possible, the valence stability of Co is lower than that of Fe. Divalent cobalt may also arise from the thermally induced charge disproportionation of Co^{3+} into Co^{4+} and Co^{2+} as reported for related oxides[35].

The introduction of Sm on the A-site of BSCF composing "BSSCF" clearly increases the electrical conductivity which is attributed to the increased concentration of electronic charge carriers because of the reduction of Co^{4+} and Fe^{4+} ions. The metal-like conductivity which is often seen in the perovskites has been attributed to the overlap of transition metal d-orbitals and oxygen p-orbitals at high temperatures[35].

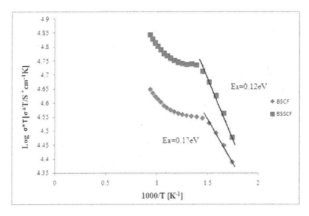

Figure 4. Conductivities for BSCF & BSSCF as 1000/T

CONSTANT COMPOSITION MEASUREMENTS

Prior to the start of the constant composition measurements, the initial stoichiometry of the determined by equilibrating in 0.21 atm at 820 ℃. Based on the thermodynamic state, the oxygen stoichiometry was determined from the literature data. Kriegel et. al have obtained oxygen stoichiometry data using the thermogravimetry measurements in combination with a redox titration and based on the thermodynamic state, the oxygen stoichiometry was determined to be 2.45[36,37] for our BSCF sample. Both $\Delta\bar{H}_{O_2}$ and $\Delta\bar{S}_{O_2}$ were determined at the fixed value x=2.45. With x fixed, the temperature was lowered at approximately 1 ℃/mi n and the emf recorded as a function of temperature. The relatively fast kinetics of the BSCF and BSSCF perovskites allowed data to be collected rapidly and data was collected after 5 min equilibration time at an interval of 5 C. Figure 5 shows the measured cell EMF as a function of temperature for both the compositions. While BSSCF data shows a linear trend, BSCF data shows an inflection around ~610 ℃. T he cells EMFs were further analyzed to determine the partial molar quantities of the materials systems. As described earlier, by converting the emf to oxygen activity, P_{O_2}, the partial molar enthalpy, $\Delta\bar{H}_{O_2}$, and partial molar entropy, $\Delta\bar{S}_{O_2}$, can be directly evaluated from the slopes and intercepts of log P_{O_2} vs 1/T plots. The BSCF exhibits two distinct regions; a high temperature region with $\Delta\bar{H}_{O_2}$ and $\Delta\bar{S}_{O_2}$ of 202 KJ/mole and 158 J/mole·K and low temperature region with $\Delta\bar{H}_{O_2}$ and $\Delta\bar{S}_{O_2}$ of 106 KJ/mole and 50 J/mole.K. On the other hand, BSSCF sample shows only one region over the entire temperature range with $\Delta\bar{H}_{O_2}$ and $\Delta\bar{S}_{O_2}$ of 149 KJ/mole and 120 J/mole·K. Two phases in BSCF compound have been reported by a few investigators and it has been speculated that the $Ba_{0.5}Sr_{0.5}Co_{0.8}Fe_{0.2}O_{3-x}$ perovskite transforms into a two phase mixture of hexagonal and cubic perovskites[23,24]. It has been suggested that the meta-stable cubic phase is formed during the fast cooling which transform to the stable hexagonal polymorph over time. The observed one slope behavior of BSSCF is indicative of single phase. Girdauskaite *el. al*[36] have used indirect method of coulometric titration and thermogravimetric methods to measure the partial molar enthalpy, $\Delta \overline{H}_{O_2}$, and partial molar entropy, $\Delta \overline{S}_{O_2}$, oxygen stoichiomery of $Ba_{0.5}Sr_{0.5}Co_{0.8}Fe_{0.2}O_{3-x}$ system for the oxygen stoichiometry of 2.55. Their reported numbers for $\Delta \overline{H}_{O_2}$, and partial molar entropy, $\Delta \overline{S}_{O_2}$, of 114 KJ/mol and 53 J/mol·K agrees well with the measured values for the high temperature cubic phase.

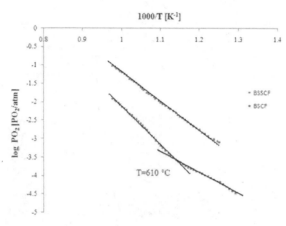

Figure 5: Log P_{O2} vs. 1000/T plots for BSCF and BSSCF perovskites using gas-tight electrochemical cells.

The correlation of the unit cell volumes with the thermodynamic data has been rationalized in terms of higher absolute values of the partial molar enthalpy, $\Delta \overline{H}_{O_2}$ associated with the higher cohesive energy resulting from the tightly bound unit cell with smaller volumes. The more stable manganates show highest absolute values of the partial molar enthalpy, $\Delta \overline{H}_{O_2}$ among perovskites and generally have lower unit cell volumes. Based on the our XRD results, BSSCF has lower unit cell volume and higher measured the partial molar enthalpy, $\Delta \overline{H}_{O_2}$, can be rationalized. Recently, a detailed HRTEM study by Elimkov et. al[23] also confirmed the transition from cubic to hexagonal phase in the cubic $Ba_{0.5}Sr_{0.5}Co_{0.8}Fe_{0.2}O_{3-x}$ perovskite[24]. It was shown that the cubic $Ba_{0.5}Sr_{0.5}Co_{0.8}Fe_{0.2}O_{3-x}$ perovskite transforms into a two phase mixture of hexagonal and cubic perovskites, which occurs via a new monoclinic perovskite. Mueller et. al[25] have also studied the the decomposition of the cubic perovskite-type oxide $Ba_xSr_{1-x}Co_{0.8}Fe_{0.2}O_{3-y}$ into hexagonal and cubic perovskite-type phases using analytical microscopes and X-Ray Diffraction (XRD). Their SEM and TEM studies indicate that the hexagonal phase grows predominantly at the grain boundaries of the cubic BSCF ceramics. Due to the sluggish kinetics, usual cooling rates lead to the meta-stable cubic BSCF, but over time the transition to the stable hexagonal polymorph may be detrimental to applications incorporating BSCF[24].

CONCLUSIONS

$Ba_{0.5}Sr_{0.5}Co_{0.8}Fe_{0.2}O_{3-x}$ (BSCF) and $(Ba_{0.5}Sr_{0.5})_{0.9}Sm_{0.1}Co_{0.8}Fe_{0.2}O_{3-x}$ (BSSCF) perovskites have been synthesized and characterized for their structure, electrical conductivity and thermodynamic behavior as a function of temperature and oxygen activity. The conductivity as a function of temperature in air for both composition shows a maxima around 420450 ℃. The thermodynamic properties, the partial molar enthalpy, $\Delta \bar{H}_{O_2}$, and partial molar entropy, $\Delta \bar{S}_{O_2}$, were measured directly using a gastight electrochemical cell for a fixed oxygen stoichiometry for both systems. The measurements show that BSCF undergoes a subtle polymorphic phase changes around 600 ℃ from cubic to hexagonal during cool down. It was found that Sm doping improves BSCF electrical conductivity and also stabilized the structure over the entire temperature range. The measured partial molar thermodynamic quantities were rationalized in terms of the unit cell volumes and phase stability. It is anticipated that the observed structural instability in the BSCF system will adversely affect the performance of devices incorporating BSCF as membrane or electrode where long-term stability of operation is required. The superior stability and electrical conductivity of Sm doped BSCF should make this materials be a good candidate for IT-SOFC cathode and ion transport membrane material.

REFERENCES
[1] Y. Teraoka, H. Zhang, S. Furukawa, N. Yamazone, Chem. Lett., 2, 1743, (1985).
[2] O. Yamamoto, Y. Takeda, R. Kanno, and M. Noda, Perovskite-type oxides as oxygen electrodes for high temperature oxide fuel cells, Solid State Ionics, 22, 241-246, (1987).
[3] J. Richter, P. Holtappels, T. Graule, T. Nakamura and L. J. Gauckler, Materials design for perovskite SOFC cathodes, Monatsh Chem, 140, 985–999, (2009).
[4] J. Yoo, A. K. Verma and A.J. Jacobson, Proc. Electrochem. Soc., PV2001-28, 27-31 (2002).
[5] B. C. H. Steele, Ceramic ion conducting membranes, Curr. Opin. Solid State Mater. Sci,, 1, 684, (1996)
[6] H. J. M. Bouwmeester and A. J. Burggraaf, Dense Ceramic Membranes for Oxygen Separation, The CRC Handbook of Solid State Electrochemistry. Edited by P.J. Gellings and H.J.M. Bouwmeester. CRC, Boca Raton, U.S.A.,, 481-553, (1997).
[7] N. Q. Minh and T. Takahashi, Science and Technology of Ceramic Fuel Cells. Elsevier, Amsterdam, (1995).
[8] C. Xia, W. L. Rauch, F. Chen, and M. Liu, "$Sm_{0.5}Sr_{0.5}CoO_3$ cathodes for low-temperature SOFCs", Solid State Ionics, 149, 11-19, (2002).
[9] L. M. Tai, M.M. Nasrallan and H.U. Anderson Solid State Chem., 118,117, (1995).
[10] Z. Shao and S. M. Haile, A high-performance cathode for the next generation of solid-oxide fuel cells, NATURE, 431, 170-173, (2004).
[11] Wei Zhou, Ran Ran, Zongping Shao, Progress in understanding and development of $Ba_{0.5}Sr_{0.5}Co_{0.8}Fe_{0.2}O_{3-x}$ based cathodes for intermediate-temperature solid-oxide fuel cells: A review, Journal of Power Sources, 192, 231–246, (2009).
[12] H. Wang, Y. Cong, W. Yang, Investigation on the partial oxidation of methane to syngas in a tubular $Ba_{0.5}Sr_{0.5}Co_{0.8}Fe_{0.2}O_{3-x}$ membrane reactor, Catalysis Today, 82, 157-166, (2003).
[13] H. H. Wang, C. Tablet, A. Feldhoff and H. Caro, Investigation of phase structure, sintering, and permeability of perovskite-type $Ba_{0.5}Sr_{0.5}Co_{0.8}Fe_{0.2}O_{3-\delta}$ membranes, J. Membr. Sci., 262, 20-26, (2005).
[13] S. B. Adler, J.A. Lane and B.C.H. Steele. J. Electrochem. Soc., 143, 3554, (1996).
[14] M. Liu, J. Winnick, Fundamental issues in modeling of mixed ionic-electronic conductors (MIECs), Solid State Ionics, 118, 11-21, (1999).
[15] J. Fleig, SOLID OXIDE FUEL CELL CATHODES: Polarization Mechanisms and Modeling of the Electrochemical Performance, Annu. Rev. Mater. Res., 33, 361-82, (2003).
[16] A. J. Jacobson, Materials for Solid Oxide Fuel Cells, Chem. Mater., 22 (3), pp 660–674 (2010)

[17] P. Ried, E. Bucher, W. Preis, W. Sitte and P. Holtappels, Characterisation of $La_{0.6}Sr_{0.4}Co_{0.2}Fe_{0.8}O_{3-\delta}$ and $Ba_{0.5}Sr_{0.5}Co_{0.8}Fe_{0.2}O_{3-\delta}$ as cathode materials for the application in intermediate temperature fuel cells, ECS Trans., **7**, 1217-1224, (2007).

[18] Z. Chen, R. Ran, Z. Shao, H. Yu, J. C. D. Costa, and S. Liu, Further performance improvement of $Ba_{0.5}Sr_{0.5}Co_{0.8}Fe_{0.2}O_{3-x}$ perovskite membranes for air separation, Ceramics International, **35**, 2455-2461, (2009).

[19] K. Ahn, H. Kim, Y, Chung, J.W. Son, H. W. Lee, J. H. Lee, Catalytic characteristics of perovskite-type oxides under mixed methane and oxygen gases, Journal of the Korean Ceramic Society, **45**, 232-237, (2008).

[20] L. Wang, R. Merkle, J. Maier, T. Acartük and U. Starke, Oxygen tracer diffusion in dense $Ba_{0.5}Sr_{0.5}Co_{0.8}Fe_{0.2}O_{3-x}$ films, Applied Physics Letters, **94**, (2009).

[21] E. Girdauskaite, H. Ullmann, V.V. Vashook, U. Guth, G.B. Caraman, E. Bucher and W. Sitte, Oxygen transport properties of $Ba_{0.5}Sr_{0.5}Co_{0.8}Fe_{0.2}O_{3-x}$ and $Ca_{0.5}Sr_{0.5}Mn_{0.8}Fe_{0.2}O_{3-x}$ obtained from permeation and conductivity relaxation experiments, Solid State Ionics, **179**, 11-12, (2008).

[22] E. Bucher, A. Egger, P. Ried, W. Sitte, and P. Holtappels, Oxygen nonstoichiometry and exchange kinetics of $Ba_{0.5}Sr_{0.5}Co_{0.8}Fe_{0.2}O_{3-x}$, Solid State Ionics, **179**, 1032-1035, (2008).

[23] K. Efimov, Q. Xu and A. Feldhoff, Transmission electron microscopy study of $Ba_{0.5}Sr_{0.5}Co_{0.8}Fe_{0.2}O_{3-\delta}$ -δ perovskite decomposition at intermediate temperatures, Chemistry of Materials, **22**, 5866-5875, (2010).

[24] S. Šarcová , K. Wiik, J. Tolchard, H.J.M. Bouwmeester, T. Grande, T. Structural instability of cubic perovskite $Ba_xSr_{1-x}Co_{1-y}Fe_yO_{3-\delta}$, Solid State Ionics, **178**, 1787-1791, (2008).

[25] D. N. Mueller, R. A. De Souza, T. E. Weirich, D. Roehrens, J. Mayer and M. Martin, A kinetic study of the decomposition of the cubic perovskite-type oxide $Ba_xSr_{1-x}Co_{0.8}Fe_{0.2}O_{3-x}$ (BSCF) (x = 0.1 and 0.5), Physical Chemistry Chemical Physics, **12**, 10320-10328, (2010).

[26] S. Li, Z. LüX. Huang, B. Wei, W. Su, Electrical and thermal properties of $(Ba_{0.5}Sr_{0.5})_xSm_{1-x}Co_{0.8}Fe_{0.2}O_{3-x}$ perovskite oxides, Solid State Ionics, **178**, 417–422, (2007).

[27] S. Li, Z. Lu, N. Ai, K. Chen, W. Su, Electrochemical performance of $(Ba_{0.5}Sr_{0.5})_{0.9}Sm_{0.1}Co_{0.8}Fe_{0.2}O_{3-x}$ as an intermediate temperature solid oxide fuel cell cathode, Journal of Power Sources **165**, 97–101, (2007).

[28] R. Hui, C. Sun, S. Yick, C. Decè s-Petit, X. Zhang, R. Maric and D. Ghosh, $Ba_{x}Pr_xCo_{1-y}Fe_yO_{3-\delta}$ as cathode materials for low temperature solid oxide fuel cells, Electrochimica Acta, **55**, 4772-4775, (2010).

[29] P. C. Young, A. Jacobson, Electrical Conductivity and Oxygen Nonstoichiometry of $La_{0.2}Sr_{0.8}Fe_{0.55}Ti_{0.45}O_{3-d}$, Journal of the Electrochemical Society, **152**, 65-73, (2005).

[30] P. C. Young, A. Frank, A. J. Jacobson, J The oxygen nonstoichiometry and electrical conductivity of $La_{0.7}Sr_{0.3}Cu_{0.2}Fe_{0.8}O_{3-d}$, Journal of Materials Chemistry, **16**, 3624-3628, (2006).

[31] J. -H. Park, R. N. Blumenthal, and M. A. Panhans, Direct Measurement of Thermodynamic Properties in Nonstoichiometric CaO-Doped Cerium Dioxide by Coulometric Titration, J. Electrochem. Soc., **135**, 855-859, (1988).

[32] J.-I. Jung, S. T. Misture and D. D. Edwards, The electronic conductivity of $Ba_{0.5}Sr_{0.5}Co_xFe_{1-x}O_{3-\delta}$ (BSCF: x = 0 ~ 1.0) under different oxygen partial pressures, Journal of Electroceramics, **24**, 261-269, (2010).

[33] G. S. Jiang , C. L. Song, D. C. Li, S. J. Feng, L. Wei, C. S. Chen, Chin. J. Chem. Phys., 17,75, (2004).

[34] B. Wei, Z. Lu, X. Huang, J. Miao, X. Sha, X. Xin, W. J. Su, Eur. Ceram. Soc., **26**, 2827-2832, (2006).

[35] Z. Yá ng, A. S. Harvey, A. Infortuna, J. Schoonman, and L. J. Gauckler, Electrical conductivity and defect chemistry of $Ba_xSr_{1-x}Co_{1-y}Fe_yO_{3-\delta}$ perovskites , Journal of Solid State Electrochemistry, (2010).

[36] E. Girdauskaite , H. Ullmann, M. Daroukh , V. Vashook, M. Blow and U. Guth, Oxygen stoichiometry, unit cell volume, and thermodynamic quantities of perovskite-type oxides, J Solid State Electrochem, **11**, 469-477, (2007).

[37] R. Kriegel, R. Kircheisen, J. Töfer, Oxygen stoichiometry and expansion behavior of $Ba_{0.5}Sr_{0.5}Co_{0.8}Fe_{0.2}O_{3-x}$, Solid State Ionics, **181**, 64-70, (2010).

ROLE OF SINTERING ATMOSPHERE ON THE STABILITY OF LSM-YSZ COMPOSITE

Manoj Mahapatra and Prabhakar Singh
Center for Clean Energy Engineering, Department of Chemical, Materials and Biomolecular
Engineering, University of Connecticut, Storrs, CT 06269, USA

ABSTRACT

We have investigated the structural and chemical stability of $La_{0.8}Sr_{0.2}MnO_3$ (LSM)-8 mol% yttria stabilized zirconia (YSZ) composite used as the cathode (air electrode) in the solid oxide fuel cell (SOFC). LSM and YSZ powders were mixed and sintered at 1400°C for 10 h in controlled atmosphere ($pO_2 = 0.21$ to 10^{-6} atm). Densification, microstructure, and reaction products formation were studied using scanning electron microscopy (SEM) and X-Ray diffraction (XRD) techniques. Rhombohedral LSM and cubic YSZ are present with reduction in pO_2 from 0.21 atm to 10^{-6} atm. Chemical reaction associated with the formation of $La_2Zr_2O_7$ and MnO_x (Mn_3O_4) compounds lowers the stability of the LSM-YSZ composite with reduced oxygen partial pressure (10^{-6} atm). Subsequent sintering in 0.21 atm pO_2 led to the disappearance of the $La_2Zr_2O_7$ and MnO_x compounds. Possible mechanisms responsible for chemical reactions and morphological changes are discussed.

1. INTRODUCTION

Solid oxide fuel cell (SOFC) is one of the most attractive electrochemical devices to cleanly and efficiently convert chemical energy into electrical energy[1-3]. For practical applications, SOFCs should demonstrate a life time of ~5000 h for mobile applications and ~ 40,000 h for stationary applications under the nominal operating conditions (650-1000°C) and exposure to oxidizing and reducing atmospheres)[1-3]. While the efficiency of SOFCs can be increased to more than 50% by exploring new materials and designs, the performance degradation with time remains one of the major challenges[4]. Instability of bulk materials, surface and interfaces at the cell and stack level (cathode, electrolyte, anode, interconnect, and gas seal) during cell fabrication and operation contribute to the electrical performance degradation.

Strontium doped lanthanum manganite (LSM), 8 mol% yttrium doped zirconia (YSZ), and nickel-YSZ cermet are most commonly utilized as SOFC cathode, electrolyte, and anode respectively[2]. These materials meet the primary requirements such as high electrical conductivity of the cathode (~200 S/cm), high ionic conductivity of the electrolyte (~0.01 S/cm), and matched thermal expansion coefficient (CTE, 10.8-11.8×10^{-6}/°C)[2, 3]. In order to improve the SOFC performance, mixed ionic electronic conductors (MIEC) as cathode materials have been investigated[5, 6]. Strontium doped lanthanum ferrite, cobaltite, and cobaltite ferrite are the examples of MIECs. These materials have high electrical and ionic conductivity, and catalytic activity for the oxygen reduction. The higher CTE (12.2-20.5×10^{-6}/°C) of these cathode materials, however, leads to the delamination of cathodes from electrolyte during SOFC operation and degrades the performance[5, 7]. It is also noted that the above properties of these materials vary with temperature and the chemical stability remains poor. For example, the conductivity of LSF ($La_{0.5}Sr_{0.5}FeO_3$) decreases from 352 S/cm at 550°C to ~180 S/cm at 900°C and reacts with YSZ above 800°C to form an insulating strontium zirconate ($SrZrO_3$) phase at the electrode–electrolyte interface[6,8]. Therefore, these materials may not be suitable for long term operation. Despite the lower ionic conductivity (5.97×10^{-7} S/cm), LSM is preferred for its superior stability to those of the MIECs and similar CTE to that of YSZ[6, 9-11]. Although the ionic conductivity and electro-catalytic activity of LSM cathode can be improved by introducing YSZ in the composite form[12], reactions between LSM and YSZ at elevated temperatures (cell fabrication and cell operation) result in the formation of lanthanum zirconate ($La_2Zr_2O_7$) and strontium zirconate ($SrZrO_3$) phases at the LSM-YSZ interface[9,13,14]. The conductivity of the $La_2Zr_2O_7$ and $SrZrO_3$ phases are in the range of 10^{-4} -10^{-5} S/cm which is much lower than those of LSM and YSZ resulting in an increase in cell resistance[9, 15].

The reaction between LSM and YSZ has been found to be predominantly dependent on the composition (Sr concentration and A/B ratio) and surface area of the starting powders, sintering temperature, exposure time, and atmosphere (oxygen pressure)[13,15-19]. A-site deficient LSM, however, enhances its densification and decreases the catalytic activity[20, 21].

Oxygen partial pressure at the LSM-YSZ interface changes during SOFC operation due to mainly four reasons: 1) the specific resistance of LSM and YSZ, 2) the tortuosity in the cathode, 3) the adsorption and desorption rates of oxygen ions at the electrochemically active area (triple phase boundary), and 4) localized temperature distribution and hot spots[22]. Changes in the oxygen partial pressure across the interface are schematically shown in Figure 1. Based on the literature survey, it is noted that limited information exists on the role of exposure atmosphere on the stability[19]. This paper describes the role of oxygen pressure on microstructural evolution, reaction kinetics and stability of LSM-YSZ composite.

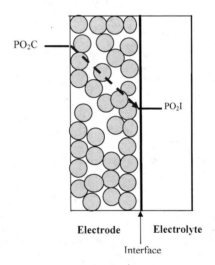

PO$_2$C

PO$_2$I

Electrode   **Electrolyte**

Interface

Figure 1 Schematic of oxygen partial pressure difference at the LSM-YSZ interface from the bulk LSM and YSZ. pO$_2$I, pO$_2$C, are the oxygen pressure at the bulk cathode, and cathode-YSZ interface.

2. EXPERIMENTAL

2.1. Sample preparation

LSM (surface area 4.66 m^2/g, FuelCell Materials, Ohio), YSZ (surface area 4.7m^2/g), and carbon black powders (density 1.8 g/cm^3, CABOT Corporation, Massachusetts) in 1:1:0.12 mass ratio were ball milled for 24 h using ZrO_2 balls and ethanol. The mixed powder batch was dried for 24 h and pressed into pellets. The LSM-YSZ pellets were bisque fired in air at 1000°C for 2 h in order to burn out carbon. The bisque fired samples were subsequently sintered at 1400°C for 10 h in flowing dry air (pO_2 ~0.21 atm) and N_2 (pO_2 ~10^{-6} atm) in a tube furnace (Model No. 0300334, CM Inc. New Jersey). Selected samples, initially sintered in N_2 atmosphere, were resintered at 1400°C for 10 h in dry air. For all the sintering experiments, the heating and cooling rates were 3°C/min. It should be noted that carbon black powder serves as pore former. The porosity of the bisque fired sample is maintained at ~ 30 vol% to uniformly expose the samples to the experimental environments.

2.2. Characterization

The bulk densities of the LSM-YSZ samples were measured by Archimedes method following ASTM C20-97 standard. The densities of at least five samples for each sintering atmospheres were measured. Average density is reported.

The LSM-YSZ samples were polished to optical finish. The polished samples were cleaned in an ultrasonic bath with water and dried and wiped with ethanol. The microstructures of the polished samples were analyzed using scanning electron microscope (SEM, Quanta 600 FEI Company, Hillsboro, OR). The EDS module (EDAX) attached to the SEM was used to analyze the elemental composition of the sintered samples. A minimum of ten EDS spot analyses was carried out for each phase. The average elemental composition is reported.

X-ray diffraction (XRD) studies, using a Bruker D8 Advance diffractometer, were carried out to identify the structure and phases present in the sintered LSM –YSZ composite. The scan step was 0.04° with CuK$_\alpha$ radiation (λ =1.5406 Å).

3. RESULTS

3.1. Density

The average densities of the sintered samples are listed in Table 1. The sintered samples showed higher density in air compared to that in nitrogen. It is also noted that the density of samples exposed to N_2 followed by air reaches similar density (~94%) to that of air sintered samples.

Table 1 Density of the LSM-YSZ composites sintered in different atmospheres

Sintering atmosphere	Density(g/cm^3)	Theoretical density (%)
Air	5.86 ± 0.04	94
N_2	5.77 ± 0.09	90
N_2 followed by air	5.88 ± 0.09	94

3.2. Phase evolution

The effect of sintering atmosphere on the compound formation and/or structural changes in the LSM-YSZ composites, as identified by XRD, is shown in Figure 2. Cubic zirconia (JCPDS number: 00-030-1468) is observed for all the samples regardless of sintering atmosphere. Rhombohedral LSM (JCPDS number: 00-053-0058) and cubic $La_2Zr_2O_7$ (JCPDS number: 00-017-0450) are observed for the samples sintered in air atmosphere. Rhombohedral (JCPDS number: 00-074-8264) LSM, cubic $La_2Zr_2O_7$ (JCPDS number: 00-017-0450), and MnO_x (Mn_3O_4) phase are found for the samples sintered N_2 atmosphere. Although rhombohedral LSM is observed for the N_2 sintered samples, the shift in XRD pattern indicates that the lattice parameters are different from that of air sintered samples leading to higher unit cell volume. Peak intensity from the XRD pattern indicates that the amount of the $La_2Zr_2O_7$ is much higher for the samples sintered in N_2 atmosphere (Figure 2). The samples sintered in N_2 followed by air atmosphere show identical XRD peak positions and intensity to those of air sintered samples indicating similar phase formation. These observations indicate that sintering atmosphere significantly affects the interaction between LSM and YSZ, the structure of LSM; and the reaction between LSM and YSZ is reversible. The same density of the samples sintered in air and N_2 followed by air atmosphere (Table 1) also supports the reversibility of LSM and YSZ interaction.

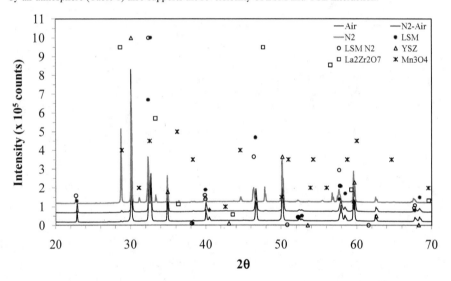

Figure 2 XRD patterns of the LSM-YSZ composites in different sintering atmospheres.

3.3. Microstructure

The microstructure of the polished LSM-YSZ composites sintered in different atmospheres is shown in Figure 3. For the sample sintered in air (Figure 3a), two phases are homogeneously distributed. Tiny pores (<1 μm) are randomly distributed at the grain boundary triple junction. The grain sizes vary from 1.5 μm to 6 μm. The bright phases are LSM (marked as 1) and the dark phases are YSZ (marked as 2) as confirmed by EDS spot analyses given in Table 2. The EDS analyses show that ~ 7 at% La and ~ 10 at% Mn diffuse into the YSZ. No $La_2Zr_2O_7$ phase is observed although it is found in XRD analysis.

For the sample sintered in N_2 atmosphere, four distinct phases are observed (Figure 3b). The grey colored phase (marked as 1) is LSM; dark phase (marked as 2) is YSZ; the brighter phase (marked as 3) is $La_2Zr_2O_7$ and the darkest phase (marked as 4) is Mn-rich as confirmed by EDS spot analyses (Table 2). The LSM and YSZ are homogeneously distributed. Mn-rich phase of ~ 10 μm in size (Mn_3O_4 as confirmed by XRD) is randomly distributed (Figure 3b). Small pores (<1 μm) are randomly distributed at the grain boundary triple junction. The grain sizes of LSM, YSZ, and $La_2Zr_2O_7$ vary from 4-6 μm. Overall, the grain sizes of the LSM and YSZ phases are larger than those of the samples sintered in air. The EDS analyses show that ~ 7.5 at% La and ~ 15 at% Mn diffuse into the YSZ.

For the samples sintered in N_2 followed by air, significant morphological changes are observed. The bright phase is LSM and a darker phase is YSZ. The grain size of the LSM and YSZ phases vary from 1-8 μm. Overall, the YSZ grains are larger than LSM. The EDS analyses show that ~ 7 at% La and ~ 12 at% Mn diffuse into the YSZ.

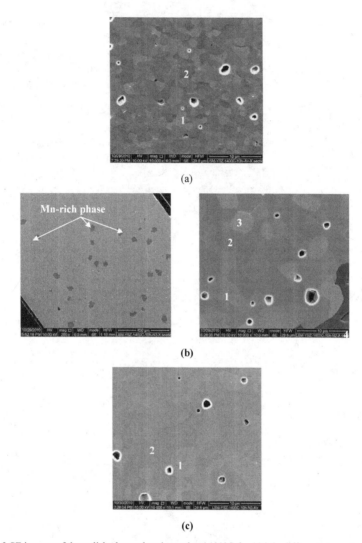

Figure 3 SE images of the polished samples sintered at 1400°C for 10 h in different atmospheres: a) air, b) N₂, and c) N₂ followed by air.

Table 2 Elemental composition of the observed phases in the LSM-YSZ composites sintered in different atmospheres (Figure 3)

Samples	La	Zr	Sr	Mn	Compounds
Air					
1	42.3 ±0.9		12.5±0.4	45.1±0.9	LSM
2	6.5±1.0	84.0±1.7		9.5±1.1	Solid solution ZrO_2 with La and Mn
N_2					
1	41.5±1.1	2.8±1.0	13.2±0.5	42.5±1.1	LSM
2	7.5±0.6	77.6±1.3		15.0±0.8	Solid solution ZrO_2 with La and Mn
3	45.9±0.1	54.1±0.1			$La_2Zr_2O_7$
N_2 followed by air					
1	41.9±2.2	2.4±2.5	11.7±1.6	44.0±1.9	LSM
2	5.8±1.0	82.1±1.4		12.0±1.1	Solid solution ZrO_2 with La and Mn

4. DISCUSSION

The effect of sintering atmosphere on the stability of the LSM-YSZ composite will be discussed from several aspects: structural and morphological changes, reaction kinetics, and the reversibility of compound formation.

Respective phase formation during the sintering in various gas atmospheres is shown in Table 2. It is observed that lanthanum and manganese from LSM diffuse into the adjacent YSZ phase due to favorable ionic radii whereas strontium ion does not diffuse due to its larger ionic radii[23]. Manganese ions diffuse readily in the YSZ due to its smaller ionic radii. The change in the lattice parameters of LSM structure is explained based on the changes in the valence state of manganese ions (Mn^{4+} to Mn^{3+}). In N_2 atmosphere ($pO_2 \sim 10^{-6}$atm), manganese ions reduce to lower valence state according to

$$\frac{1}{2}O_2 + V_O^{\cdot\cdot} + 2Mn'_{Mn} = 2Mn^{\cdot}_{Mn} \qquad (1)$$

$$\frac{1}{2}O_2 + V_O^{\cdot\cdot} + 2Mn'_{Mn} = 2Mn^x_{Mn} + O_O^x \qquad (2)$$

where $V_O^{\cdot\cdot}$, Mn^x_{Mn}, $Mn^{\cdot}_{Mn}$, and Mn'_{Mn} are oxygen vacancy, Mn^{4+}, Mn^{3+}, and Mn^{2+}, respectively. Reduction of Mn^{4+} and Mn^{3+} ions with larger Mn^{2+} ions increases the unit cell volume of the LSM. Subsequently, the LSM lattice parameters are changed in N_2 atmosphere leading to peak shifts in the XRD pattern. As the Mn^{2+} ion is more soluble into the YSZ than those of Mn^{3+} and Mn^{4+} ions due to the favorable ionic radii, the concentration of manganese ion in the YSZ is the highest for the samples sintered in N_2 atmosphere[24]. Further sintering in air (pO_2 =0.21 atm) oxidizes the Mn^{2+} ions into the Mn^{3+} ions. As a result, the solubility of manganese ions into the YSZ decreases as supported by the EDS spot analyses presented in Table 2. The unit cell volume of the LSM also decreases as supported by the reappearance of same structure[25].

The overall grain size of the LSM-YSZ composites sintered in N_2 atmosphere is larger than those for the samples sintered in air. In N_2 atmosphere, oxygen vacancies increase due to change in the oxidation state of manganese ions as stated above. Also the oxygen vacancies increase according to

$$O_O^x = V_O^{\cdot\cdot} + 2e' + \frac{1}{2}O_2 \qquad (3)$$

The oxygen vacancies move to the grain boundaries and free surfaces with the concentration of oxygen vacancies highest in the neck regions[26, 27]. The concentration gradient of oxygen vacancies enhances the atomic mobility towards the neck regions decreases the surface energy resulting in larger grain size than those of sintered in air[26]. For the samples sintered in N_2 atmosphere followed by air, the grain size varied from 1-8 μm likely due to the higher sintering time.

$La_2Zr_2O_7$ phase forms due to reaction between the LSM and YSZ and is influenced by the thermodynamic and kinetic limitations including chemical potential, Gibbs free energy, temperature, and oxygen partial pressure. Kinetic limitations include elemental inter-diffusion, reaction rate and time. According to La_2O_3-MnOx-ZrO_2 chemical potential diagram, lanthanum manganite and zirconia cannot be in equilibrium without $La_2Zr_2O_7$ at high temperatures according to the reaction[28, 29]:

$$LaMnO_3 + ZrO_2 = \frac{1}{2} + La_2Zr_2O_7 + MnO_x\frac{3-x}{2} + O_2 \qquad (4)$$

The forward reaction rate of the above reaction increases (k) with decreasing pO_2 (k α $pO_2^{-(1.5-x/2)}$).Subsequently, $La_2Zr_2O_7$ formation is accelerated in N_2 atmosphere. Also, the Gibbs free energy of $La_2Zr_2O_7$ formation is the lowest (~ 20kJ/mol) if MnO_x co-exists as Mn_3O_4[28]. In the present experimental conditions, Mn_3O_4 is the most stable oxide phase among all other form of manganese oxides (MnO_2, Mn_2O_3, MnO, and Mn_3O_4)[30]. Accordingly, Mn_3O_4 is observed for the samples sintered in N_2 atmosphere. ZrO_2 excess nonstoichiometric $La_2Zr_2O_7$ is more stable than stoichiometric and/or La_2O_3 excess $La_2Zr_2O_7$ as observed in EDS spot analyses (Table 2)[31].

For the air sintered sample, $La_2Zr_2O_7$ and Mn_3O_4 phases are not observed during the SEM analysis although very small amount of $La_2Zr_2O_7$ has been detected using XRD analysis. This can be explained from reaction kinetics. In LSM-YSZ composite, manganese diffuses faster into the YSZ than that of lanthanum resulting in La_2O_3 excess LSM at the LSM-YSZ interface. The excess La_2O_3 reacts with ZrO_2 at the LSM-YSZ interface and forms $La_2Zr_2O_7$. The LSM at the LSM-YSZ is less La_2O_3 excess in air that that in N_2 atmosphere. Accordingly, very small amount of $La_2Zr_2O_7$ and Mn_3O_4 form. $La_2Zr_2O_7$ may have formed in the nm region at the LSM-YSZ interface, the SEM resolution limit[19].

For the samples sintered in N_2 atmosphere followed by air, the quantity of $La_2Zr_2O_7$ is significantly smaller compared to that of N_2 and almost similar to that of air sintered sample as supported by $La_2Zr_2O_7$ peak intensity in XRD pattern (Figure 2). This indicates that the reaction products formed have the tendency to revert back to initial products according to the reaction:

$$\frac{1}{2} + La_2Zr_2O_7 + \frac{1}{3}Mn_3O_4 + \frac{4}{3}O_2 = LaMnO_3 + ZrO_2 \qquad (5)$$

Further investigations are needed to understand the LSM-YSZ reaction reversibility.

5. CONCLUSION

The role of sintering atmosphere on the structural and chemical stability of $La_{0.8}Sr_{0.2}MnO_3$ (LSM)-8 mol% yttrium stabilized zirconia (YSZ) composite has been studied as a function of oxygen partial pressure. On decreasing pO_2 from 0.21 atm to 10^{-6} atm, the unit cell volume increases due to change in the oxidation state of manganese. At higher pO_2 the LSM and YSZ remain stable whereas at lower pO_2, LSM and YSZ preferentially react to form additional $La_2Zr_2O_7$ and Mn_3O_4 phases. Reversibility of above reactions has been demonstrated.

ACKNOWLEDGEMENT

The authors sincerely acknowledge the financial support from Siemens Corporation under the grant number 635383. Technical discussion with Dr. S.D. Vora, NETL, is acknowledged.

REFERENCES
[1] Fuel Cell Hand Book, 7th edition, EG&G Technical Services, U.S. Department of Energy, Office of Fossil energy, National Energy Technology Laboratory, Morgantown, WV, 2004.
[2] M. C.Williams, Solid Oxide Fuel Cells: Fundamentals to Systems, Fuel Cells, 7, 78-85 (2007).
[3] A. J. Jacobson, Materials for Solid Oxide Fuel Cells, Chem. Mater., 22, 660-74 (2010).
[4] M. Yokoo, Y. Tabata, Y. Yoshida, K. Hayashi, Y. Nozaki, K. Nozawa, and H. Arai, Highly Efficient and Durable Anode-Supported SOFC Stack with Internal Manifold Structure, J. Power Sources, 178, 59-63 (2008).
[5] C. Sun, R. Hui, and J. Roller, Cathode Materials for Solid Oxide Fuel Cells: A review, J. Solid State Electrochem., 14, 1125-44, (2010).
[6] E. V. Bongio, H. Black, F. C. Raszewski, D. Edwards, C. J. McConville, and V. R. W. Amarakoon, Microstructural and High-Temperature Electrical Chracterization of $La_{1-x} Sr_x FeO_{3-\delta}$, J. Electroceram., 14, 193-98, (2005).
[7] A. Petric, P. Huang, and F. Tietz, Evaluation of La–Sr–Co–Fe–O perovskites for Solid Oxide Fuel Cells and Gas Separation Membranes, Solid State Ionics, 135, 719-25 (2000).
[8] A. M. Amesti, A. Larrañaga, L. M. R. Martínez, A. T. Aguayo, J. L. Pizarro, M. L. Nó, A. Laresgoiti, M. I. Arriortua, Reactivity Between La(Sr)FeO3 Cathode, Doped CeO2 Interlayer and Yttria-stabilized zirconia Electrolyte for Solid Oxide Fuel Cell Applications, J. Power Sources, 185, 401-10 (2008).
[9] S. P. Jiang, Development of Lanthanum Strontium Manganite Perovskite Cathode Materials for Solid Oxide Fuel Cells: A Review, J. Mater. Sci., 43, 6799-6833 (2008).
[10] J. Mizusakia, Y. Yonemurab, H. Kamatab, K. Ohyamab, N. Mori, H. Takai, H. Tagawa, M. Dokiya, K. Naraya, T. Sasamoto, H. Inaba, and T. Hashimoto, Electronic Conductivity, Seebeck Coefficient, Defect and Electronic Structure of Nonstoichiometric $La_{1-x} Sr_x MnO_3$, Solid State Ionics, 132, 167-80 (2000).
[11] S.P. Jiang, A Comparison of O_2 Reduction Reactions on Porous (La,Sr)MnO3 and (La,Sr)(Co,Fe)O3 Electrodes, Solid State Ionics. 146, 1-22 (2002).
[12] J. M. Vohs and R. J. Gorte, High-Performance SOFC Cathodes Prepared by Infiltration, Adv. Mater., 21, 943–956 (2009).
[13] A. Mitterdorfer and L. J. Gaukler, $La_2Zr_2O_7$ Formation and Oxygen Reduction Kinetics of the $La_{0.85}Sr_{0.15} Mn_yO_3$, O2 (g)/YSZ System, Solid State Ionics, 111, 185-218 (1998).
[14] M. Backhaus-Ricoult, Interface Chemistry in LSM-YSZ Composite SOFC Cathode, Solid State Ionics, 177, 2195-2200 (2006).

[15]F. W. Poulsen and N. V. Puil, Phase Relations and Conductivity of Sr- and La- Zirconates, Solid State Ionics, **53-56**, 777-83 (1992).

[16]G. Stochoniol, E. Syskakis, and A. Naoumidis, Chemical Compatibility between Strontium-Doped Lanthanum Manganite and Yittria-Stabilized Zirconia, J. Am. Ceram. Soc., **78**, 929-32 (1995).

[17]J. A. M. Roosmalen and E. H. P. Cordfunke, Chemical Reactivity and Interdiffusion of (La, Sr)MnO$_3$ and (Zr, Y)O$_2$, Solid Oxide Fuel Cell Cathode and Electrolyte Materials, Solid State Ionics, **52**, 303-12 (1992).

[18]M. Chen, Y. L. Liu, A. Hagen, P. V. Hendriksen, and F. W. Poulsen, LSM-YSZ Reactions in Different Atmospheres, Fuel Cells, **9**, 833-40 (2009).

[19]A. Chen, G. Bourne, K. Siebein, R. DeHoff, E. Wachsman, and K. Jones, Characterization of Lanthanum Zirconate Formation at the A-site-Deficient Strontium-Doped Lanthanum Manganite Cathode/Yittrium-Stabilized Zirconia Electrolyte Interface of Solid Oxide Fuel Cells, J. Am. Ceram. Soc., **91**, 2670-75 (2008).

[20]J. W. Stevenson, P. F. Hallman, T. R. Armstrong, and L. A. Chick, Sintering Behavior of Doped Lanthanum and Yttrium Manganite, J. Am. Ceram. Soc., **78**, 507-12 (1995).

[21]D. L. Meixner and R. A. Cutler, Sintering and Mechanical Characteristics of Lanthanum Strontium Manganite, Solid State Ionics, **146**, 273-84 (2002).

[22]A. V. Virkar, A Model for Solid Oxide Fuel Cell (SOFC) Stack Degradation, J. Power Sources, **172**, 713-24 (2007).

[23]C. C. T. Yang, W. C. J. Wei, and A. Roosen, Reaction Kinetics and Mechanisms Between La$_{0.65}$Sr$_{0.3}$MnO$_3$ and 8 mol% Yttria-Stabilized Zirconia, J. Am. Ceram. Soc., **87**, 1110-16 (2004).

[24]T. Kawashima, Phase Equilibria in ZrO$_2$-Y$_2$O$_3$-MnO$_t$ Ternary System at 1673K, Mater. Trans., **40**, 967-71 (1999).

[25]E. K. Andersen, I. G. K. Andersen, P. Norby, and J. C. Hanson, Kinetics of Oxidation of Fuel Cell Cathode Materials Lanthanum Strontium Manganates (III)(IV) at Actual Working Conditions: In Situ Powder Diffraction Studies, J. Solid State Chem., **141**, 235-40 (1998).

[26]M. N. Rahaman, Ceramic Processing and Sintering, Mercel Dekker Inc., New York (1995).

[27]B. P. McCarthy, L. R. Pederson, H. U. Anderson, X. D. Zhou, P. Singh, G. W. Coffey, and E. C. Thomsen, Enhanced Shrinkage of Lanthanum Strontium Manganite (La$_{0.90}$Sr$_{0.10}$MnO$_{3+\delta}$) Resulting from Thermal and Oxygen Partial Pressure Cycling, J. Am. Ceram. Soc., **90**, 3255-62 (2007).

[28]M. Chen, Understanding the Thermodynamics at the LaMnO$_3$-YSZ Interface in SOFC, PhD thesis, Swiss Federal Institute of Technology, Zurich, 2005.

[29]M. Chen, A. N. Grundy, B. Hallstedt, and L. J. Gaukler, Thermodynamic Modelling of the La-Mn-Y-Zr-O System, CALPHAD, **30**, 489-500 (2006).

[30]S. Fritsch and A. Navrotsky, Thermodynamic Properties of Manganese Oxides, J. Am. Ceram. Soc., **79**, 1761-68 (1996).

[31]C. R. Stanek, L. Minervini, and R. B. Grimes, Nonstoichoimetry in A$_2$B$_2$O$_7$ Pyrochlores, J. Am. Ceram. Soc., **85**, 2792-98 (2002).

Interconnects

CROFER22 APU IN REAL SOFC STACKS

Qingping Fang, Mario Heinrich, Christian Wunderlich
staxera GmbH.
Dresden, Saxony, Germany

ABSTRACT

Among the potential ferritic candidates for SOFC applications, Crofer22 APU has been proven to be one of the best materials considering its high oxidation resistance. Even so, bare Crofer22 APU is still not sufficient to fulfill the lifetime requirement of 40,000 hours at relatively high working temperature (~850°C) of ESC stacks. The lifetime of Crofer22 APU can be extended, without lowering the operating temperature, by different protective coatings. Our previous work has shown that spinel coatings such as MnCoFe (MCF) and CuNiMn (CNM) can effectively retard the Cr evaporation and increase the durability of SOFC stacks.

In order to compare the protective effects of MCF and CNM coatings on the real Crofer22 APU bipolar plates under real operating conditions, two short stacks with mixed coatings were built and tested. The stacks contained both uncoated layers as well as layers with MCF and CNM coating. The protective layers were also prepared by different coating techniques, such as roll-coating (RC), high velocity oxygen fuel spraying (HVOF) and galvanic plating. Post mortem analysis (PMA) was performed after stack testing. Bipolar plate cross sections were investigated by SEM/EDX and the Cr content within the cathodes was determined by ICP-OES, which showed that MCF coating was more effective in Cr retarding than CNM coating.

The MCF coated Crofer22 APU bipolar plates and cells from other separated stacks after long term operation were also analyzed by SEM/EDX and ICP-OES. Based on the obtained results, especially the oxide growth rate, a lifetime prediction of 0.5mm Crofer22 APU sheet under SOFC working conditions is given.

INTRODUCTION

In electrolyte supported cell (ESC) based SOFC systems, the relatively high working temperature (≥800°C), together with the demands on low electrical resistance, high corrosion resistance, and suitable thermal expansion, limit the selection of possible metallic materials as a SOFC interconnector. High chromium ferritic stainless steels have been intensively investigated and are being commonly used in SOFC due to their proper physical and chemical properties.[1-4] Among those materials, Crofer22 APU, developed by Forschungszentrum Jülich and commercialized by ThyssenKrupp VDM, is specially designed for SOFC applications.[5-7] Nevertheless, bare Crofer22 APU, especially the thin metal sheet, is still not sufficient to fulfill requirements such as low Cr evaporation and good oxidation resistance, considering the lifetime target of more than 40,000 hours. Cr evaporation from chromia forming alloys, including Crofer22 APU, inside stacks may lead to two types of degradation:

i. Oxidation and corrosion of the interconnector will increase the contact resistance to other stack components, and in the worst case, may also cause breakaway oxidation;

ii. Deposition of vaporized Cr^{6+} species in the cathode function layer, i.e. cathode poisoning, degrades the cell performance.

In order to minimize the Cr evaporation from Crofer22 APU, different coating approaches based on perovskite or spinel compounds have been developed and evaluated.[8-11] Our previous works have shown that spinel coatings such as $MnCo_{2-x}Fe_xO_4$ (MCF) and $Cu_xNi_{1-x}Mn_2O_4$ (CNM) can effectively retard the Cr migration of Crofer22 APU, and decrease the growing rate of the oxide scales.[12,13]

The function of the protective coatings are generally investigated in laboratory scale by cyclic oxidation and mass gain using small model samples under ambient air or simulated anode atmospheres, respectively. In this work, real stacks were built with both coated and uncoated Crofer22 APU bipolar plates, and tested in dual atmosphere conditions. The function of the coatings, as well as the oxidation behavior of Crofer22 APU, can be reviewed first from the difference in the cell performance, and later by further investigations in terms of oxide scale morphology, and Cr content in steel and cathode. Based on the obtained testing results, the lifetime of 0.5mm Crofer22 APU (coated and uncoated) sheet at 850°C is predicted using a simplified model. As a preliminary evaluation, the effect of the oxide scale stress and delamination were not taken into consideration in this work for the lifetime prediction.

EXPERIMENTS
Stack preparation
A standard staxera stack design consists of 30 cells. In order to minimize the influence of the temperature gradient along the stack height on the coating performance, short stacks consisting of 10 cells were built. For the same reason, the furnace operation mode, which has better temperature homogeneity, was chosen instead of the normal hot box operation mode.

Coatings of $MnCo_{1.9}Fe_{0.1}O_4$ (MCF) and $Cu_{0.6}Ni_{0.4}Mn_2O_4$ (CNM) were chosen for comparison. For roll coating (RC), pastes of MCF and CNM were prepared by the Fraunhofer Institute for Ceramic Technologies and Systems (IKTS) using commercially available powders. The high velocity oxygen fuel spraying (HVOF) of MCF powder was done by the Fraunhofer Institute for Material and Beam Technology (IWS). Additionally, Ni and Cu layer were tried to be galvanic plated separately on Crofer22 APU, with the expectation that Cu, Ni and Mn from Crofer22 APU could form CNM spinel intrinsically during the stack operation

Table I shows the construction of the two 10-cell stacks. In each stack, all the stack components, except for the difference in coatings on the bipolar plates, were all out of the same batch, so that the deviation of the cell performance in each stack would mainly due to the difference in coatings.

Table I. Construction of the two 10-cell stacks

Cell No.	Bipolar plate (stack 1)	Bipolar plate (stack 2)
10	uncoated	uncoated
9	uncoated	uncoated
8	RC MCF	RC MCF
7	RC MCF	RC MCF
6	HVOF MCF	RC CNM
5	HVOF MCF	RC CNM
4	uncoated	uncoated
3	uncoated	uncoated
2	RC CNM	Galvanic Ni and Cu
1	RC CNM	Galvanic Ni and Cu

Stack testing
As previously mentioned, the 10-cell stacks were tested in a furnace operation mode. The furnace temperature was adjusted to avoid the highest temperature inside the stack exceeding 860°C. The stack temperature was monitored by thermocouples at the top and bottom clamping plates, center of the stack, as well as the air inlet and outlet. The stack temperatures varied within a small range during the test, depending on the operating mode. A list of the operating parameters is shown in Table II. Since the tests were mainly focused on the comparison of the different coating functions, relatively

mild operating conditions were chosen to avoid any side effects from high fuel utilization and high current density.

Table II. Operating parameters of the two 10-cell stack tests

Operating parameters	Stack 1	Stack 2
Stack design	Mk200	Mk100
T_{max} (°C)	860	860
cathode gas	air	air
fuel composition	$50\%H_2/40\%N_2/10\%H_2O$	$40\%H_2/60\%N_2$
fuel utilization	16~40%	30~50%
operating time (h)	~4000	~3500
current density (mA/cm²)	85~273	200

Besides the two 10-cell stacks, another 5-cell short stack was built, using the Mk100 bipolar plates (cassette made of 0.5mm Crofer22 APU with HVOF coated MCF on active region) after a 13000-hour stack operation. This short stack was also operated in the furnace, which was set at 850°C. It was firstly operated under 200mA/cm² with a nominal fuel utilization of 50% for 4000 hours, and followed by 3500 hours under 123mA/cm² with a nominal fuel utilization of 75% due to a higher contact resistance from the deformed external contact sheet and a leakage through a broken MEA (membrane electrode assembly). The stack was then switched into OCV operation under purge gas ($90\%N_2/10\%H_2$) because of the leakage through another broken MEA. It must be emphasized, that the recycling of the tested bipolar plates brought difficulties and tolerance errors in building the new stack, which increased highly the risk of cell fracture. In the standard staxera Mk200 and Mk100 stacks, the risk of broken MEA is much lower than this specially prepared short stack. The bipolar plates of the 5-cell stack have been operated at working temperature under dual atmosphere for a total time of more than 26,000 hours, and no breakaway oxidation has been observed under normal working conditions, even at uncoated regions.

Stack analysis

Post mortem analysis was conducted with the 10-cell stacks after stack operation. The stacks were disassembled, and examined firstly by visual inspection. Samples of interconnector were cut from different layers for cross section preparation. Samples of MEA from different cells at similar position were taken for ICP-OES measurement. The measured Cr amount in cathodes is another evaluation of the coating function. The preparation of polished cross sections, SEM/EDX investigations and ICP-OES measurements of the 10-cell stack were all conducted by Leibniz Institute for Solid State and Materials Research (IFW).

RESULTS

Stack 1: 10-cell Mk200 stack

Fig.1 shows cell voltages of the stack and also the temperature of the top clamping plate of the stack. During the test, the bottom clamping plate is always about 10°C colder than the top one. The stack was mainly operated with a current density of 227mA/cm² (29A) and fuel utilization of 40%. Different load and fuel utilization are also marked in the figure. All cells with uncoated interconnector (cell 3, 4, 9 and 10) show much worse performance compared to other cells. Fluctuation of the cell voltages (cell 3, 4 and 9) was later proven by post mortem analysis to be due to the delamination of the cathode contact ribs from the interconnector, which means that the coating layers improve the adhesion of contact ribs and interconnector.

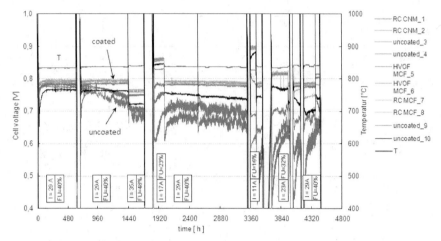

Figure 1. Cell voltages of stack 1.

The degradation rates of cell power and area specific resistance (ASR) were calculated and plotted in Fig.2 and Fig.3, respectively. Data at t_0=400h and t=4400h were taken for the calculation. The highest degradation rates of the cells with uncoated bipolar plates (cell 3, 4 and 9) were later proven to be mainly due to the poor adhesion between cathode contact ribs and bipolar plates. Nevertheless, the degradation rate of cell 10 (with uncoated bipolar plate, too) is still the highest one among the rest of the cells. No obvious difference can be seen from all the other coated layers, except cell 7 and 8, where the relatively higher degradation rates were again due to a fracture of the cell 7 shortly before ending of the test.

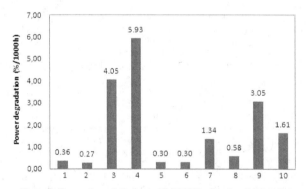

Figure 2. Power degradation rate (%/1000h) of each cell in stack 1.

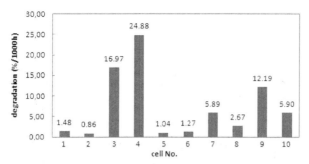

Figure 3. Area specific resistance (ASR) degradation rate (%/1000h) of each cell in stack 1.

The amount of Cr in the cathode layer was measured by ICP-OES. Table III shows the results of ICP-OES analysis. Without coatings, cell 4 and 10 show the highest Cr concentration in the cathode. With MCF coating, either roll coating or HVOF spraying, the Cr concentrations are nearly 10 times lower than that of uncoated one, and close to the detecting limit of the instrument. Despite the similar degradation rates, the Cr content in CNM coated cell is quite higher compared to MCF coated ones. Under current testing conditions, the cell degradation is not directly related to the amount of Cr in the cathode.

Table III. Cr in the cathodes in stack 1 (measured by ICP-OES).

Cell No.	Coating	Cr (μg/cm^2)	Cr (wt%)
10	uncoated	22.4	2.21
8	RC MCF	3.0	0.30
5	HVOF MCF	4.5	0.44
4	uncoated	34.6	3.42
2	RC CNM	11.8	1.16

Fig.4 shows the variation of Cr concentration from air side (left) to fuel side (right) in uncoated and RC MCF coated bipolar plate. Without coating, a clear concentration gradient can be seen inside the interconnector, with the concentration near air side being the lowest due to the higher Cr evaporation rate under high oxygen partial pressure. With RC MCF coating, Cr is more homogeneously distributed through the interconnector, with the concentration being close to the original value of ~22%. The concentration line of Cr from HVOF MCF and RC CNM coated interconnector shows the similar behavior to RC MCF coated one.

Fig.5 presents the SEM image of the cross section of bare Crofer22 APU interconnector at the air side. The typical double-layer oxide scale structure consists of a thicker Cr_2O_3 sub layer and a thin $MnCr_2O_4$ spinel layer on top. The thickness of the inner Cr_2O_3 layer is in the range of 8~22μm.

Fig.6 shows the SEM images of the cross section and also the compositional analysis of the interface of Crofer22 APU with RC and HVOF coated MCF coating at air side. Fe was missing during the elemental line scan, thus the composition of the bulk steel, except Cr concentration, are not correct. For RC coated MCF, the porous coating layer is well bonded to the Cr_2O_3 layer through a dense oxide layer, which consists of Mn, Co, Fe and Cr and was formed during the stack operation. Evaporation of Cr from the steel substrate was well hindered by this dense oxide layer. Compared to roll coating, the formation of the dense layer in between coating and substrate is not obvious with HVOF spraying, since the complete coating is more dense. There is a crack inside the dense HVOF coated MCF layer, which is close to the interface of Cr_2O_3 and coating. Whether the crack was formed during stack operation or sample preparation is still not fully clear. The white dots in the images of HVOF coated MCF are particles from the cathode contact ribs, which could be brought during the sample preparation.

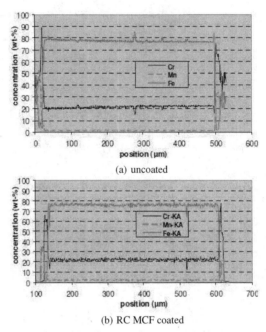

(a) uncoated

(b) RC MCF coated

Figure 4. Line scan profile of Cr in (a) uncoated and (b) RC MCF coated interconnector.

Figure 5. SEM image of the cross section of bare Crofer22 APU at air side.

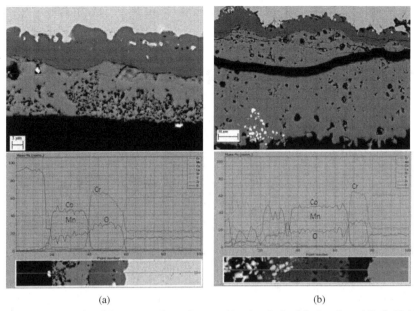

(a)   (b)

Figure 6. SEM images of the cross section and compositional analysis of the interface of Crofer22 APU and (a) RC and (b) HVOF coated MCF coating at air side.

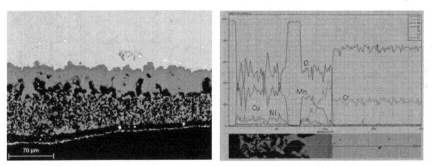

Figure 7. SEM images of the cross section and compositional analysis of the interface of Crofer22 APU and RC CNM coating.

The SEM images of the cross section and compositional analysis of the interface of Crofer22 APU and RC CNM coating is shown in Fig. 7. Similar to RC MCF coating, the porous CNM layer is well bonded to Cr_2O_3 through a dense oxide layer, which consists of Cu, Ni, Mn and Cr. The amount of Cr in this dense layer is higher compared to the one with RC MCF coating. The relatively large difference in the thickness of the coating (including the dense oxide layer) between RC MCF

(12~15µm) and RC CNM (70~90 µm) was due to the higher viscosity of the CNM paste during coating process.

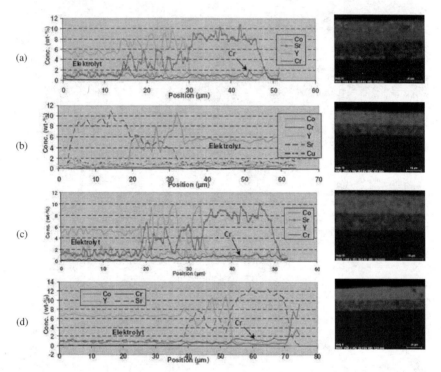

Figure 8. Line scan and Cr mapping of cathode of the cell with (a) uncoated, (b) RC MCF, (c) HVOF MCF and (d) RC CNM coating at air channel positions

Cr poisoning of the cathode is one of the serious degradation mechanisms for SOFC using high chromium ferritic interconnectors. Elemental line scan and mapping were used to identify preliminarily the amount and distribution of Cr species inside the cathode. Fig.8 presents the results of line scan and Cr mapping at air channel positions. No Cr was detected in the sample with RC MCF coating. A slight drop of Cr concentration from cathode contact layer to function layer can be seen in other three samples. Possible deposition of Cr at the electrolyte/cathode interface was detected only in uncoated cell by both line scan and mapping.

The cathodes in contacting with the cathode contact ribs were also investigated in the same way. The results are similar, but the accumulation of Cr at the interface of electrolyte and cathode were not observed in any of the invested samples.

Stack 2: 10-cell Mk100 stack

The 10-cell Mk100 stack was first operated in a constant current mode for about 2500 hours, and followed by thermal cycles. The degradation rate after the stationary operation was calculated with the measured data at t_0=500h and t=2500h. Fig.9 plots the averaged cell voltages of each two-cell block. The cells with Ni and Cu electroplated interconnectors show large voltage drop at the early stage of the test, which was attributed to the oxidation of the metallic coating layers. The cells with either MCF or CNM coated interconnectors show stable performance, while those with uncoated interconnectors have already shown clear tendency of degradation. The power and ASR degradation rates of the cells are plotted in Fig.10 and Fig.11, respectively. The difference in degradation rates between the two uncoated cell blocks is again proven to be due to the poor adhesion of the cathode contact ribs.

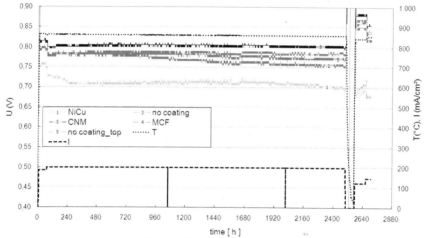

Figure 9. Cell voltages of stack 2.

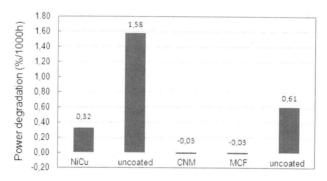

Figure 10. Power degradation rate (%/1000h) of each 2-cell block in stack 2.

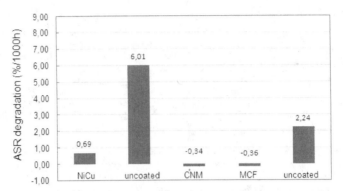

Figure 11. ASR degradation rate (%/1000h) of each 2-cell block in stack 2.

The Cr amount inside the cathodes was measured by ICP-OES, and the results are shown in Table IV. Combined with the results in Table III, it can be easily concluded that MCF coating shows the best protective function against Cr evaporation among the tested coatings.

Table IV. Cr content inside the cathodes in stack 2 (ICP-OES).

Cell No.	Coating	Cr ($\mu g/cm^2$)	Cr (wt%)
9	uncoated	18.2	1.80
8	RC MCF	1.8	0.18
5	RC CNM	8.1	0.80
4	uncoated	19.0	1.88
2	Ni Cu	19.2	1.90

5-cell stack with recycled bipolar plates from long term operated stack

Staxera has demonstrated low degradation rates (<0.4%/1000h) of stacks made of coated Crofer22 APU interconnector after either long-term stationary or thermal cycling operations.[14] A 5-cell

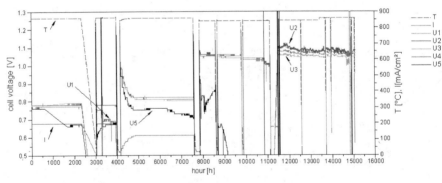

Figure 12. Cell voltages of the 5-cell stack using recycled bipolar plates.

stack of Mk100 design was built using the bipolar plates recycled from one of the long-term tested stacks, focusing the investigation on possible lifetime of the interconnector. The short stack contains bipolar plates (coated with HVOF MCF), which have been operated for 13000 hours, and also an uncoated plate for comparison. The detailed operating procedure has been described in the experimental part. Fig.12 presents the performance of the stack. The bottom cell (U1) was closed and shortcut manually after ~4000h due to the leakage or MEA fracture. The top cell (U5), which is contacting with an uncoated Crofer22 APU plate, showed a faster voltage drop between ~700h and 1700h, which could be attributed to the poor adhesion of the cathode contact ribs from our previous experience. The faster degradation of the top cell from 6400h was mainly the result of a leakage or MEA fracture. All other cells show stable voltages during the operation. The test station was changed at around 11000h, which could be the reason to the slight fluctuation of all cell voltages after restarting.

Fig. 13 shows two pictures of the stack after ~13,000h operation. The bipolar plates have been operated under a dual atmosphere at ~850°C for a total time of about 26,000 hours. As shown in Fig.13 (a), breakaway oxidation can be seen at the corner where there was a burning through a leakage of the bottom layer. Other part of the stack, even without MCF coating, didn't show breakaway oxidation as shown in Fig.13 (b).

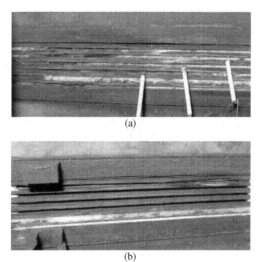

(a)

(b)

Figure 13. 5-cell stack after 13,000h furnace operation at 850°C. (a) Breakaway oxidation due to burning through a leakage of the bottom layer; (b) no breakaway oxidation under normal working conditions even at uncoated regions.

30-cell stacks with 13000h operation time

Besides the short stack tests mentioned above, standard 30-cell stacks, both Mk100 and Mk200 designs, have been subjected to long term stationary tests for nearly 13000h. The stacks were analyzed in terms of oxide scale morphology, and Cr content in steel and cathode. The results were compared to stacks with different operating time, and the observed dependence of oxide growth rate on operating time is used below for prediction of the lifetime of 0.5mm Crofer22 APU at 850°C.

Prediction of the lifetime of 0.5mm Crofer22 APU at 850°C

When the lifetime of Crofer22 APU is considered to be only limited by oxidation behavior, then depletion of Cr content in Crofer22 APU (from 22% to 18%) due to the growth and spallation of the protective oxide scale (i.e. Cr_2O_3 and $MnCr_2O_4$ for bare Crofer22 APU) and Cr evaporation after long time operation will lead to breakaway oxidation. The lifetime is then in turn limited by the Cr evaporation rate, scale growth rate and adhesion/spallation of the oxide scales. Based mainly on the experimental results, a simplified model, taking into consideration of the Cr evaporation, oxide scale growth from coated and uncoated Crofer22 APU was used to estimate the lifetime of 0.5mm Crofer22 APU under stack operating conditions. As a preliminary estimation, the effect of scale adhesion/spallation was not taken into consideration.

Assuming Cr evaporation is completely blocked by the protective coating, and then the maximum thickness of the Cr_2O_3 scale can be calculated with the following assumptions:

1) All the Cr species out of the bulk Crofer22 APU form Cr_2O_3 scale at the surface
2) Outward diffusion of Cr through Cr_2O_3 is the rate limiting process, and Cr inside the bulk material is homogeneously distributed.
3) Breakaway oxidation takes place when the Cr concentration drops to 18%. Our previous investigation has shown that the Cr concentration is in a range of 14~16% near the regions, where breakaway oxidation occurs. To avoid overestimation of the Cr_2O_3 thickness, 18% is used in the calculation.

Given a same oxide growing speed in both air and fuel sides, which is not the real case but safer for the calculation of the lifetime, a maximum Cr_2O_3 thickness of 26µm at air side is calculated. If the oxide growth rate in air is considered to be twice of that of fuel side, then a maximum thickness of 35µm at air side is obtained. The calculated maximum Cr_2O_3 thickness will be compared with the testing results, and used for the prediction of the lifetime of 0.5mm Crofer22 APU according to a parabolic law.

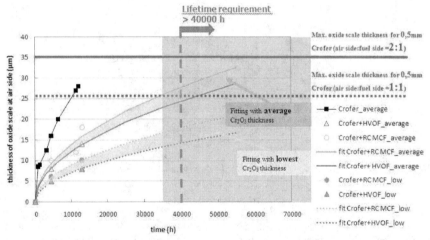

Figure 14. Prediction of the lifetime of 0.5mm Crofer22 APU by parabolic fitting of the measured thickness of Cr_2O_3 layer after model testing and stack testing.

The thicknesses of Cr_2O_3 layers from the tested 10-cell and 30-cell stacks are plotted in Fig.14. The measured minimum thickness of the Cr_2O_3 layers from RC MCF and HVOF MCF coated bipolar plates are marked by filled circle and triangle, respectively. The calculated average thicknesses of the Cr_2O_3 layers are then marked by empty circle and triangle, respectively. Only the results from MCF coated bipolar plates are plotted out of the all coated samples, since MCF coating gives the best results concerning both Cr evaporation and oxide scale growth. The measured data points were fitted using parabolic relationship. The solid and dotted horizontal lines present the maximum thickness of Cr_2O_3 layer of 0.5mm Crofer22 APU sheet at air side under the assumptions that Cr_2O_3 growth rate at fuel side is half and same as that of air side, respectively. The intersection of the fitted lines and the horizontal lines indicate roughly the lifetime of MCF coated 0.5mm Crofer22 APU interconnector, which is more than 35,000 hours assuming a same oxide growth rate in air and fuel gas using the fitting lines of average thickness of Cr_2O_3 layers. Since the real Cr_2O_3 growth rate at air side is faster than at fuel side under current testing conditions, the calculated maximum Cr_2O_3 thickness are actually underestimated. Thus the deduced lifetime is also underestimated. As a comparison to MCF coated Crofer22 APU, the measured values from uncoated plates were also plotted in Fig.14.

CONCLUSIONS

Two 10-cell stacks consisting of RC MCF, HVOF MCF, RC CNM and (Ni, Cu) coated together with uncoated bipolar plates made of 0.5mm Crofer22 APU sheet were operated with dual atmospheres at 850°C. Due to the initial oxidation process, electroplated Ni and Cu introduced additional resistance to the stack, and thus are not suitable for the protective function. All other spinel coatings improved the adhesion of cathode contact ribs on the interconnector and decreased the oxide growth and Cr evaporation rates. The coated cells, except the electroplated ones, show the low degradation rate of ~0.3%/1000h under constant current mode, while the best uncoated cell has a degradation rate of nearly 1%/1000h. Despite the similar degradation rate of CNM and MCF coated cells, a higher Cr evaporation rate was observed from RC CNM coated cell by measuring the Cr content inside cathodes. A shorter lifetime from RC CNM coated interconnector compared to MCF coated one is thus expected concerning the higher Cr evaporation rate. There is no large difference in protective functions of MCF coating by roll coating and HVOF spraying. With a thicker and denser coating layer by HVOF, the Cr_2O_3 growth rate is lower compared to roll coating, where a porous coating layer was obtained. But a thicker and denser coating layer may have the disadvantages of less tolerance to the stress during the oxide growth due to different expansion coefficients. The crack along the interface of coating and Cr_2O_3 scale in case of HVOF MCF coating could be such an indication. This could also explain why there was more Cr inside cathode with HVOF MCF coating, since Cr could move out easily along the cracks. The lifetime of the interconnector depends not only on the oxide growth rate, but also on the adherence of the oxide scales and the protective coating. In this case, a thick and dense protective coating layer may not be favorable in view of the high risk of spallation and delamination.

The 5-cell stack built with the recycled bipolar plates from one of the long-term operated stacks shows stable performance. The bipolar plates, with HVOF MCF coating on the active region, didn't show breakaway oxidation under normal working conditions after a total operation time of 26,000 hours, even at uncoated region.

Based on the experimental results, the lifetime of MCF coated 0.5mm Crofer22 APU sheet was predicted using parabolic fitting. The effect of scale spallation was not considered at the preliminary stage. According to the current calculation, a lifetime of 40,000 hours is achievable. Longer lifetime could also be expected by further improving the coating properties and qualities.

REFERENCES
[1]F. Tietz, H.-P. Buchkremer, D. Stöver, Components manufacturing for solid oxide fuel cells, *Solid State Ionics*, **152-153**, 373-381 (2002).
[2]W.J. Quadakkers, J. Piron-Abellan, V. Shemet, L. Singheiser, Metallic interconnectors for solid oxide fuel cells – a review, *Materials at High Temperatures*, **20(2)**, 115-127 (2003).
[3]P. Kofstad, R. Bredesen, High temperature corrosion in SOFC environments, *Solid State Ionics*, **52**, 69-75 (1992).
[4]P. Huczkowski, W.J. Quadakkers, Effect of geometry and composition of Cr steels on oxide scale properties relevant for interconnector applications in Solid Oxide Fuel Cells (SOFCs), PhD thesis, (2007), http://www.fz-juelich.de/zb/juwel
[5]W. J. Quadakkers, T. Malkow, J. Pirón-Abellán, U. Flesch, V. Shemet, and L. Singheiser, Suitability of Ferritic Steels for Application as Construction Materials for SOFC Interconnects, *Proc. Fourth European Solid Oxide Fuel Cell Forum*, 827 (2000).
[6]J. Pirón-Abellán, V. Shemet, F. Tietz, L. Singheiser, W. J. Quadakkers, and A.Gil, Ferritic Steel Interconnect for Reduced Temperature SOFC, Proc. Solid Oxide Fuel Cells VII, H. Yokokawa and S. C. Singhal, eds., *The Electrochem.Soc.*, 811 (2001).
[7]Frederick S. Pettit and Gerald H. Meier, Fundamental Studies of the Durability of Materials for Interconnects in Solid Oxide Fuel Cells, DOE report, OSTI ID:833895, (2003).
[8]W.J. Quadakkers, H. Greiner, M. Hänsel, A. Pattanaik, A.S. Khanna, W. Malléner, Compatibility of perovskite contact layers between cathode and metallic interconnector plates, *Solid State Ionics*, **91**, 55 (1996).
[9]K. Fujita, K. Ogasawara, Y. Matsuzaki, T. Sakurai, Prevention of SOFC cathode degradation in contact with Cr-containing alloy, *J. Pow. Sources*, **131**, 261 (2004).
[10]M. Stanislowski, J. Frojtzheim, L. Niewolak, W.J. Quadakkers, K. Hilpert, T. Markus, L. Singheiser, Reduction of chromium vaporization from SOFC interconnectors by highly effective coatings, *J. Pow. Sources.*, **164**, 578 (2007).
[11]N. Orlovskaya, A. Coratolo, C. Johnson, R. Gemmer, Structural characterization of lanthanum chromite perovskite coating deposited by magnetron sputtering on an iron based chromium-containing alloy as a promising interconnect material for SOFCs, *J. Am. Cer. Soc.*, **87**, 1981 (2004).
[12]V. Sauchuk, M. Kusnezoff, N. Trofimenko, S. Megel, H.P. Baldus, A. Reinert, Development of Effective Protective Materials for SOFC metallic interconnects, *Proc. 8th European Solid Oxide Fuel Cell Forum*, (2008).
[13]Q.P. Fang, M. Heinrich, C. Wunderlich, Crofer22 APU as a SOFC interconnector material, *Proc. 9th European Solid Oxide Fuel Cell Forum.* (2010).
[14]Staxera company presentations, http://www.staxera.de/Presentations.769.0.html

ASSESSMENT OF CHROMIUM EVAPORATION FROM CHROMIA AND ALUMINA FORMING ALLOYS

Sanjit Bhowmick, Gavin Le, Atul Verma and Prabhakar Singh
Center for Clean Energy Engineering
Department of Chemical, Materials and Biomolecular Engineering
University of Connecticut, Storrs, CT 06269, USA

ABSTRACT

Chromium evaporation and oxide scale growth from selected chromia and alumina forming alloys (Aluchrom YHf, 602 CA and AISI 310S) have been studied in the humidified air environment. The alloy samples were oxidized at 850 °C and 950 °C for 500 h in flowing air containing 2.6% and 12% water vapor. The evaporated Cr species from the alloys were condensed, collected and analyzed using ICP-MS. Oxide scale and metal–oxide interface morphology along with scale composition were examined using SEM, FIB and EDS. A correlation between Cr evaporation rates and oxide scale morphology is presented and discussed.

INTRODUCTION

Heat exchangers are the critical components of power generation systems of solid oxide fuel cell (SOFC) that efficiently manage the thermal energy distribution and utilization in the balance of plant (BOP) [1-5]. The advancement of fuel cell technology and thermal management system has encouraged the improvement of the heat exchanger materials. Since most of the solid oxide fuel cells operate at temperature 600 °C and higher, efficient thermal management is one of main issues for economical operations for fuel cells. Different types of low cost Fe- and Ni-based alloys are widely used to meet the requirements of heat exchangers. However, because of higher operating temperature and humid environment in the cathodic air, most of the materials are not suitable for such applications. For example, above 700 °C the commonly used alloys are susceptible to severe oxidation and creep deformation [6, 7]. The presence of water vapor accelerates the rate of oxidation and increases Cr evaporation [8-13]. The stability and the performance of the alloys are influenced by several factors such as microstructure, grain size, chemical composition, phase precipitation, scale formation, and operating temperature.

Most of the commercially available alloys used in the thermal management and interconnect applications of SOFC are contributing to the serious challenges of Cr poisoning of the air electrodes [6, 14-16]. At an evaluated temperature, the evaporated Cr species from the alloy surfaces migrate and interact with the cathode materials [17, 18]. The interaction of Cr with cathode leads to the Cr substitution into perovskite lattice and the formation of respective chromate, Cr_2O_3, and spinel ($CrMn_2O_4$) at the surface and triple phase boundaries [19, 20]. The formation of such secondary phases results in the rapid deterioration of the oxygen reduction rate which reduces the cell performance significantly. The materials degradation also reduces the overall electrical efficiency of the cell and that hinders the widespread commercialization of the SOFC technology.

The overall objective of the on-going study in our laboratory is to develop chemically and structurally stable heat exchanger materials that can withstand desired service lifetime at elevated temperature of 700 °C–950 °C. The candidate materials must remain cost effective and should offer excellent oxidation resistance and high temperature strength in the presence of water vapor. In this article, selected Ni-base and Fe-base alumina and chromia forming alloys, Aluchrom YHf, 602 CA and AISI 310S, are examined in the simulated environment of SOFC. The elemental compositions of the alloys are shown in Table 1. Aluchrom YHf, a ferritic chromium steel with addition of more than 5% aluminum, exhibits good oxidation resistance under isothermal and cyclic conditions, but displays poor mechanical properties at elevated temperatures. Austenitic stainless steel 310S are typically used for elevated temperature applications (1150°C). High chromium and nickel contents in AISI 310S provide corrosion resistance and maintain high-temperature mechanical properties. Ni-base alloy 602 CA exhibits excellent high-temperature creep properties and fatigue strength in the high cycle and low cycle fatigue modes. In this study, Cr evaporation rates of the alloys were determined as a function of temperature and gas composition, and the results were used to estimate the degradation rate of an SOFC unit. The structural and chemical characteristics of the scale after Cr evaporation tests were studied to understand the effectiveness for surface protection.

EXPERIMENTAL

Cr Evaporation Experiments

Cr evaporation experiments were conducted using the transpiration method in which the volatile Cr species formed on the heated samples were carried away using gas flow and collected in a condenser. An experimental setup was assembled for the measurement of chromium evaporation rates from the selected alumina and chromia forming alloys, Aluchrom, 602CA and AISI 310S. Figure 1 shows a schematic of the experimental setup which includes furnace assembly, gas flow system and reactor along with a condenser. A room temperature water bubble humidification was designed and fabricated to create an environment with water vapor pressure of 2-4%. To simulate an environment with higher water content (> 5%), a flash evaporator was designed and assembled, which is also equipped with high accuracy positive displacement pumps (accuracy 0.1 nano-liter/hr of liquid water). An air-flow of 300 sccm was monitored during all the experiments using a mass-flow controller.

The alloy samples were cut in 25 mm x 25 mm size coupons and polished using 600 grid and 800 grid SiC polishing papers. The samples were ultrasonically cleaned and degreased using ethanol and water. The Cr evaporation experiments were carried out at 850 ^{0}C and 950 ^{0}C for 500 h with heating and cooling rates of 5 °C/min. After Cr-evaporation experiments, Cr-species dissolved in condensed distilled water was collected. To collect Cr species deposited on the reaction tube, the tube was cut in pieces, immersed in the solution of alkaline potassium permanganate ($KMnO_4$) and sodium hydroxide (NaOH) at 80 ^{0}C, and kept for 1 h. At the end, the pieces of the tube were also rinsed and cleaned using known amount of distilled water. All the solution were added together, and 100 ml of the sample was analyzed to determine the amount of Cr by inductively coupled plasma mass spectrometry (ICP-MS).

Characterization of the Oxide Scale

The morphology and composition of the oxide scale were studied using an environmental scanning electron microscopy (ESEM). The cross-sections of the scale were prepared using a focused ion beam (FIB) instrument (FEI strata 400S; 30 keV Ga+ ions) with a dual-beam column, combining a scanning electron beam and an ion beam in one unit. The dual beam allows high-resolution imaging of the surface of the sample during the ion-beam milling process.

RESULTS

The Cr evaporation experiments on Aluchrom, 602 CA and AISI 310S samples were carried out at 850 ^{0}C and 950 ^{0}C in air with humidity of 2.6% and 12%. The calculated Cr evaporation rates from the alloys are presented in Figure 2. It is observed that the Cr evaporation rate from a conventional Cr forming alloy AISI 310S is approximately two orders of magnitude higher than a conventional alumina forming alloy Aluchrom at 850 °C in humidified air containing 2.6% water. Ni-base alloy 602 CA shows one order of magnitude lower Cr evaporation rate compared to AISI 310S, and one order of magnitude higher than Aluchrom. Figure 2 also compares Cr evaporation rates of Aluchrom and 602 CA samples as a function of water vapor pressure and temperature. For 602 CA, Cr evaporation rate is found approximately five times higher when the temperature increases from 850 °C to 950 °C. From Figure 2, it is also observed that Cr evaporation rate significantly increases with increasing water vapor pressure for Aluchrom samples. The Cr evaporation rate increases 3-5 times when the water vapor changes from 2.6% to 12%.

The surface morphology and cross-section of the scale developed on Aluchrom after 500 h Cr evaporation tests at 850 °C and 950 °C are shown in Figure 3. The presence of transitional alumina particles can be observed in the scale developed at 850 °C in 2.6% and 12% water as shown in Figure 3a and 3c. The scale formed in 2.6% water is found to be more uniform and dense compared to 12% water. The sample exposed to 12% water at 850 °C reveals overgrowth and very thin regions. The thicker scales is ~3.5 m, whereas the thickness in the thinner region is ~0.4 m at 850 °C in 12% water. In 2.6% water, a maximum scale thickness of ~2.9 m is observed at 850 °C. Figure 3e and 3f display oxide scale morphologies developed on Aluchrom at 950 °C in 12% water. X-ray diffraction studies confirmed the formation of stable alpha alumina in the scale. Scale thickness varies between ~3.9 – 0.9 m at 950 °C in 12% water. The scale developed in 12% water reveals higher porosity than the scale developed in 2.6% water. The amount of porosity in the scale also increases with increasing temperature

The surface morphology and cross-section of the scale developed on the alloy 602 CA after 500 h Cr evaporation tests at 950 °C in 2.6% water vapor pressure show localized overgrowth regions as well as thin scale formation (Figure 4). A low magnification image (Figure 4a) displays nodular growth of the oxide scale on the thin regions. Higher magnification image in Figure 4b shows the morphology of the overgrowth region. The cross-sectional images of the scale are shown in Figures 4c and 4d. The oxide thickness in the overgrowth region is ~10 m, whereas the thickness of the thinner region is ~500 nm. Two different contrasts can be observed in the thinner region of the scale as shown in the higher magnification image in Figure 4d. A columnar morphology appears as a thicker layer in the scale at the metal-scale interface, and a more equiaxed morphology is observed as a thinner layer at the gas-scale interface. The EDS study indicates that the scale in Figure 3d is predominantly alumina, whereas the overgrowth scale in Figure 4c consists of oxides of Fe, Cr and Ni. Zr-rich and Y-rich particles were also observed in the overgrowth region. A discontinuous thin alumina scale forms at metal-overgrowth scale interfaces.

The surface and cross-section morphology of the oxide scale developed on AISI 310S after 500 h Cr evaporation tests at 850 °C in 2.6% water vapor pressure are shown in Figure 5. Continuous scale without delamination or spallation is observed on the surface. Mostly finer faceted grains cover the surface of the scale; relatively larger, overgrown faceted crystals form in localized regions as shown in Figure 5b. No significant compositional variation was found between the overgrown crystals and the smaller crystals. EDS study indicated a multi-layer structure of the scale containing spinel $(Cr,Mn)_3O_4$ on the surface followed by Cr-oxide underneath. Silica particles were also observed at the metal-scale interface. Though the surface morphology does not reveal any cracks and porosities, severe cracks and pores are observed at the alloy-scale interface as shown in Figure 5c.

DISCUSSION

In this study, Cr evaporation from selected alumina and chromia forming alloys were examined at 850 °C and 950 °C in 2.6% (p_{H2O} ~ 0.026 atm) and 12% (p_{H2O} ~ 0.12 atm) humidified air environments. The surface and cross-sectional morphology of the oxide scales after Cr evaporation experiments were characterized using SEM and FIB. A schematic in Figure 6 highlights a correlation between the various oxide scale formation on the alloy surfaces and the rate of Cr evaporation. The result indicates that Cr evaporation rate from thermally grown alumina scales on Aluchrom is approximately two orders of magnitude lower than that formed on a conventional chromia forming alloy AISI 310S. The higher Cr evaporation is due to the formation of a uniform Cr-rich scale on the surface of AISI 310S. Cr evaporation rate of a Ni-base alloy 602 CA is found to be approximately one order of magnitude lower than AISI 301S. The lower Cr evaporation rate in 602 CA is partly due to the development of thin protective alumina scale on the surface along with the formation of Cr, Fe and Ni-rich islands.

The alloys form various types of chromium containing vapor species at high temperatures in oxidizing atmospheres with/without water vapor [21-24]. The present study, consistent with literature [8, 21, 23], indicates that water vapor plays a significant role in Cr release from the alloy surfaces. An increase of water vapor pressure from 0.026 atm to 0.12 atm enhances Cr evaporation rate by approximately 3-5 times for a conventional alumina forming alloy. The formation of main volatile Cr species in oxidizing atmosphere containing water vapor can be expressed by the following reactions [25, 26].

$$Cr_2O_3 \text{ (s)} + 1.5O_2 \text{ (g)} = 2CrO_3 \text{ (g)} \tag{1}$$

$$Cr_2O_3 \text{ (s)} + H_2O \text{ (g)} = 2CrO(OH) \text{ (g)} \tag{2}$$

$$Cr_2O_3 \text{ (s)} + 1.5O_2 \text{ (g)} + 2H_2O \text{ (g)} = 2CrO_2(OH)_2 \text{ (g)} \tag{3}$$

The total pressure of Cr species can be expressed as

$$p = p(CrO_3) + p(CrO_2(OH)_2) + p(CrO(OH)_2) \tag{4}$$

Using thermodynamic computation, Gindorf et al. showed that CrO_3 and $CrO_2(OH)_2$ are the most abundant species in the humidified air at elevated temperatures [16]. Considering $p(CrO(OH)_2)$ $<< p(CrO_3)$ or $p(CrO_2(OH)_2)$,

$$p = p(CrO_3) + p(CrO_2(OH)_2) \tag{5}$$

From equations (1) and (3), it can be stated that the partial pressure of CrO_3 is independent of partial pressure of water at a constant temperature, whereas the partial pressure of $CrO_2(OH)_2$ depends on the partial pressure of water. From equation (3), it can be derived that $p(CrO_2(OH)_2)$ is proportional to $p(H2O)$. Replacing $p(CrO_2(OH)_2)$ by $p(H_2O)$ in equation (5),

$$p = p(CrO_3) + b. \, p(H_2O) \tag{6}$$

Where b is a constant.

The total Cr vapor pressure for Aluchrom sample at 850 °C and 950 °C in air with water vapor pressure of ~ 0.026 atm and 0.12 atm were calculated from the Cr evaporation rates shown in Figure 2. Although the scale developed on Aluchrom is predominantly alumina, major Cr evaporation occurs in this alloy due to the formation of Cr containing nodules in the beginning of the oxidation experiment. A minor amount of Cr may also release from the dissolve Cr in the alumina scale [28]. A variation of CrO_3 and $CrO_2(OH)_2$ partial pressure in aluchrom with water vapor pressure and temperature is presented in Table 2. The result indicates that the water vapor pressure plays an important role to control the major Cr species. It has also been reported in the literature that for a given water vapor pressure, $CrO_2(OH)_2$ dominates at the lower temperature range, and CrO_3 dominates at the higher temperature range [16, 25, 26]. A transition temperature (T_t), at which the partial pressure of CrO_3 and $CrO_2(OH)_2$ become equal ($p_{CrO3} = p_{CrO2(OH)2}$), increases with increasing water partial pressure [16]. At the low partial pressure of water, in this study $p(H_2O)$ ~ 0.026 atm, it is calculated that CrO_3 is a dominating vapor species at 850 °C and 950 °C. However, $CrO_2(OH)_2$ becomes a major vapor species at 850 °C and 950 °C when the water vapor pressure increases to ~ 0.12 atm. The morphology of the scale developed on the alloys is also influenced by p_{H2O}. In this study, the scale formed in p_{H2O} ~ 0.12 atm is found more porous than the scale developed in p_{H2O} ~ 0.026 atm. It has been reported that higher water vapor pressure increases cation diffusion in the alloys which leads to vacancy condensation resulting in the porous structure in the scale [9, 11, 27].

Recently, Stanislowski et al. have provided a performance analysis correlating the voltage degradation of an SOFC unit and Cr evaporation from alloys and coatings [14, 28]. Their analysis assumes that the voltage degradation of the cell occurs only due to Cr interaction with an LSM cathode; no other factors of degradation are considered. Stanislowski et al. determined that 1% degradation of the cell voltage occurs due to 3.96 gm/cm^2 of Cr release [14]. Based on this analysis, voltage degradation effects of the present alloys are estimated for 500 h and listed in Table 3. For Aluchrom samples, the voltage degradation increases to approximately two times when the water content in the environment changes from 2.6% to 12%. In contrast to Aluchrom, the voltage degradation rates for 602 CA and AISI 310S are found higher. The voltage degradation for 602 CA is found to be ~ 4% at 850 °C, and ~ 13% at 950 °C. A voltage degradation of 17% is predicted for AISI 310S after 500 h of operation. The degradation analysis and Cr evaporation results of the alloys indicate that Ni-base 602 CA and Fe-base AISI 310S samples may not be suitable for heat exchangers applications in SOFC. In such alloys, surface coatings can be considered for the reduction of Cr evaporation rates. The preoxidation of Aluchrom, which forms a stable oxide scale on the surface, may further reduce Cr evaporation rate and improve the cell performance. However, the mechanical properties of the alloys at elevated temperatures should also be considered for designing of a stable and high-performance heat exchanger.

CONCLUSION

In this study, Cr evaporation rates from selected Fe- and Ni-base alloys (Aluchrom YHf, 602 CA and AISI 310S) were determined using the transpiration method at 850 °C and 950 °C in air containing 2.6% and 12% water. The Cr evaporation rate from Aluchrom has been found to be approximately two orders of magnitude lower than that from a conventional chromia forming alloy AISI 310S. Cr evaporation rate of a Ni-base alloy 602 CA is determined to be approximately one order of magnitude lower than AISI 301S. The surface and cross-section morphology of the scale confirm that the higher Cr evaporation rate in AISI 310S is due to the formation of a uniform Cr-rich scale on the surface. The lower Cr evaporation rate in 602 CA is due to the formation of the scale consisting of mixed Cr, Fe and Ni-oxides along with the protective alumina scale on the surface.

ACKNOWLEDGEMENT

The project is funded by the Rolls-Royce Fuel Cell Systems (US) Inc, North Canton, OH. The discussion with Dr. Manoj Mahapatra is greatly acknowledged.

REFERENCES

[1]   P. Singh and N. Q. Minh, International Journal of Applied Ceramic Technology 1 (2004) 5.
[2]   B. C. H. Steele, Journal of Materials Science 36 (2001) 1053.
[3]   L. Magistri, A. Traverso and A. F. Massardo, Journal of Fuel Cell Science and Technology 3 (2006) 111.
[4]   S. Elangovan, S. Balagopal, J. Hartvigsen, I. Bay, D. Larsen, M. Timper and J. Pendleton, Journal of Materials Engineering and Performance 15 (2006) 445.

[5] M. R. Amin and J. D. Lindstrom, Strojniski Vestnik-Journal of Mechanical Engineering 51 (2005) 418.
[6] Z. G. Yang, M. S. Walker, P. Singh, J. W. Stevenson and T. Norby, Journal of the Electrochemical Society 151 (2004) B669.
[7] I. G. Wright and R. B. Dooley, International Materials Reviews 55 (2010) 129.
[8] Z. G. Yang, G. G. Xia, P. Singh and J. W. Stevenson, Solid State Ionics 176 (2005) 1495.
[9] M. H. B. Ani, T. Kodama, M. Ueda, K. Kawamura and T. Maruyama, Materials Transactions 50 (2009) 2656.
[10] H. Asteman, J. E. Svensson and L. G. Johansson, Oxidation of Metals 57 (2002) 193.
[11] H. Asteman, J. E. Svensson and L. G. Johansson, Journal of the Electrochemical Society 151 (2004) B141.
[12] T. Horita, H. Kshimoto, K. Yamaji, N. Sakai, Y. P. Xiong, M. E. Brito and H. Yokokawa, International Journal of Hydrogen Energy 33 (2008) 3962.
[13] N. K. Othman, N. Othman, J. Zhang and D. J. Young, Corrosion Science 51 (2009) 3039.
[14] M. Stanislowski, J. Froitzheim, L. Niewolak, W. J. Quadakkers, K. Hilpert, T. Markus and L. Singheiser, Journal of Power Sources 164 (2007) 578.
[15] N. Sakai, T. Horita, K. Yamaji, Y. P. Xiong, H. Kishimoto, M. E. Brito and H. Yokokawa, Solid State Ionics 177 (2006) 1933.
[16] C. Gindorf, L. Singheiser and K. Hilpert, Journal of Physics and Chemistry of Solids 66 (2005) 384.
[17] J. W. Fergus, International Journal of Hydrogen Energy 32 (2007) 3664.
[18] H. Yokokawa, T. Horita, N. Sakai, K. Yamaji, M. E. Brito, Y. P. Xiong and H. Kishimoto, Solid State Ionics 177 (2006) 3193.
[19] S. P. Jiang and Y. D. Zhen, Solid State Ionics 179 (2008) 1459.
[20] M. C. Tucker, H. Kurokawa, C. P. Jacobson, L. C. De Jonghe and S. J. Visco, Journal of Power Sources 160 (2006) 130.
[21] B. Pujilaksono, T. Jonsson, M. Halvarsson, I. Panas, J. E. Svensson and L. G. Johansson, Oxidation of Metals 70 (2008) 163.
[22] S. R. J. Saunders, M. Monteiro and F. Rizzo, Progress in Materials Science 53 (2008) 775.
[23] K. Segerdahl, J. E. Svensson and L. G. Johansson, Materials and Corrosion-Werkstoffe Und Korrosion 53 (2002) 479.
[24] D. J. Young and B. A. Pint, Oxidation of Metals 66 (2006) 137.
[25] E. J. Opila, D. L. Myers, N. S. Jacobson, I. M. B. Nielsen, D. F. Johnson, J. K. Olminsky and M. D. Allendorf, Journal of Physical Chemistry A 111 (2007) 1971.
[26] M. Stanislowski, E. Wessel, K. Hilpert, T. Markus and L. Singheiser, Journal of the Electrochemical Society 154 (2007) A295.
[27] H. Asteman, J. E. Svensson and L. G. Johansson, Corrosion Science 44 (2002) 2635.
[28] M. Stanislowski, E. Wessel, T. Markus, L. Singheiser and W. J. Quadakkers, Solid State Ionics 179 (2008) 2406.

Table 1: Elemental composition of Aluchrom YHf , 602 CA and AISI 310S.

	Fe	Cr	Al	Mn	S	C	Ti	Si	Y	Cu	Hf	Zr	Ni
Aluchrom	bal	21	6	0.5		0.05		0.5	0.1		0.1	0.07	0.3
602 CA	8-11	24-26	1.8-2.4	0.1		0.15-0.25	0.1-0.2	0.5	0.05-0.12	0.1		0.1	bal
AISI 301S	bal	24-26		2	0.03	0.08		1.5					19-22

Table 2: Calculated partial pressure of CrO_3 and $CrO_2(OH)_2$ from the experimental measurements of total Cr vapor pressure for Aluchrom sample at 850 °C and 950 °C.

T (°C)	$p(H_2O)$ (atm)	p total, 10^{-10} (atm)	$p(CrO_3)$, 10^{-10} (atm)	$p(CrO2(OH)2)$, 10^{-10} (atm)
850	0.026	7.2	4.9	2.3
850	0.12	15.7	4.9	10.8
950	0.026	10.0	7.6	2.4
950	0.12	18.7	7.6	11.1

Table 3: Cr release rates from the alloys and corresponding estimated voltage degradation of an SOFC unit after 500 h of operation.

Alloy	T (°C)	$p(H_2O)$	Cr release in 500 h (g/cm^2)	Voltage degradation (%) after 500 h
Aluchrom	850	0.026	0.8	0.2
Aluchrom	950	0.12	1.3	0.4
Aluchrom	850	0.026	2.4	0.6
Aluchrom	950	0.12	2.5	0.7
602 CA	850	0.026	12.6	3.6
602 CA	950	0.026	51.3	14.6
310S	850	0.026	61.4	17.5

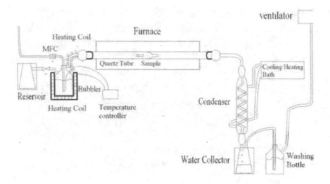

Figure 1: A schematic of Cr evaporation experimental arrangement showing the reaction tube, condenser, gas flow using humidifier and furnace.

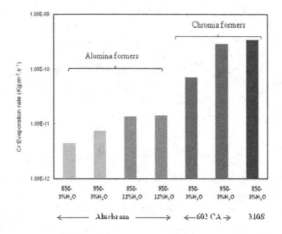

Figure 2: Cr evaporation rates of Aluchrom, 602 CA and AISI 310S as a function of temperature and water vapor pressure.

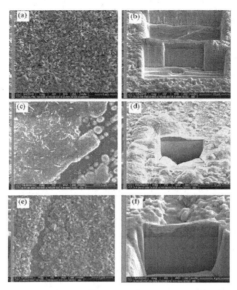

Figure 3: Surface and cross-section morphology of the scale developed on Aluchrom at (a) and (b) 850 C in 2.6% water; (c) and (d) 850 C, 12% water; and (e) and (f) 950 C, 12% water after 500 h Cr evaporation experiments.

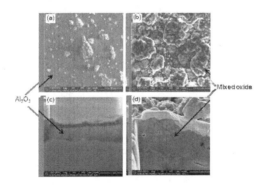

Figure 4: (a) Low magnification and (b) high magnification FESEM images of oxide scale morphology developed on 602 CA at 950 ^{0}C in 2.6% water environment after 500 h experiments. The cross-section morphology of the scale are shown in (c) thin region and (d) over growth region.

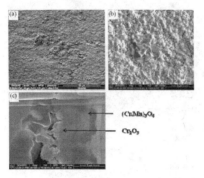

Figure 5: FESEM images of (a) and (b) Surface and (c) cross-section oxide scale morphology developed on 310 S at 850 C in 2.6% water environment after 500 h. Surface images display localized overgrown faceted crystals in the scale. The cross-section image shows multi-layer structure of the scale containing spinel $(Cr,Mn)_3O_4$ on the surface, Cr-oxide underneath and Si-oxide particles at the metal-scale interface.

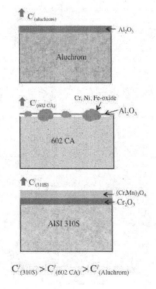

$$C'_{(310S)} > C'_{(602\ CA)} > C'_{(Aluchrom)}$$

Figure 6: A schematic showing correlation between the development of various types of oxide scales on the alloy surfaces and Cr evaporation rates. C' is the Cr evaporation rate.

EFFECT OF CHROMIUM DOPING ON THE CRYSTAL STRUCTURE, ELECTRICAL CONDUCTIVITY AND THERMAL EXPANSION OF MANGANESE COBALT SPINEL OXIDES

Yingjia Liu, Kangli Wang, Jeffrey W Fergus
Materials Research and Education Center, Auburn University
Auburn, AL, USA

ABSTRACT

Metallic interconnects have reduced the cost of solid oxide fuel cells, however their oxidation during operation and chromium volatilization can cause serious problems, such as chromium poisoning and cell degradation, so ceramic coatings have been developed as barriers to chromium and oxygen diffusion. Among them, $Mn_{1.5}Co_{1.5}O_4$ spinel has shown promising performance. To evaluate the long-term stability of the coating the properties of the reaction layer formed between the spinel coating and the Cr_2O_3 scale on the alloy surface need to be characterized. In this work, the crystal structure, electrical conductivity and thermal expansion of several manganese cobalt spinel samples, $(Mn,Co)_{3-x}Cr_xO_4$ ($0 \leq x \leq 2$) were determined. The crystal structure was determined using x-ray diffraction, electrical conductivity was measured by the 4-probe DC method and thermal expansion was tested from room temperature to 1000 °C. With increasing Cr content the cubic crystal structure was stabilized, the electrical conductivity decreased and thermal expansion coefficient decreased. The decrease in electrical conductivity indicates that formation of the reaction layer, $(Mn,Co)_{3-x}Cr_xO_4$, may lead to an increase in electrical resistance of the spinel coating, which will increase the overall cell resistance.

INTRODUCTION

Solid oxide fuel cells (SOFCs) are highly efficient and environmentally friendly power devices. The interconnect is a critical part in SOFCs because it not only connects the individual cells electrically, but also separates the air and fuel atmospheres of different cells.[1] The current research focuses on lowering the operating temperature within 600 to 800°C without compromising the materials performance. For the interconnect, metallic alloys seem to be a good choice due to their easy fabrication and low cost compared with their ceramic counterparts. Chromia forming alloys are good candidate interconnect material, but their oxidation during operation and chromium volatilization from the oxidation scale can cause serious problems, such as chromium poisoning in cathodes and cell degradation, so ceramic coatings have been developed as barriers to chromium and oxygen diffusion.[2] Among them, $Mn_{1.5}Co_{1.5}O_4$ spinel has shown promising performance.[3,4] To evaluate its long-term stability, the transport properties of the reaction layer formed between the $Mn_{1.5}Co_{1.5}O_4$ coating and the Cr_2O_3 scale on the alloy surface need to be characterized, because the interaction can change the composition of coating, and thus its properties, such as electrical conductivity and thermal expansion.

In our work, the reaction layer was found to contain two regions.[5,6] At steady state the layer in contact with Cr_2O_3, $Mn_{0.4}Co_{0.6}Cr_2O_4$ grew by diffusion of cobalt and manganese from $Mn_{1.5}Co_{1.5}O_4$. At the same time, chromium dissolved into $Mn_{1.5}Co_{1.5}O_4$ and formed the other layer. Although both layers had the spinel structure, they grew by different mechanisms. To investigate the effect of the reaction layer on the fuel cell performance, manganese cobalt spinels with different chromium content, $(Mn,Co)_{3-x}Cr_xO_4$ ($0 \leq x \leq 2$), were prepared. This work focuses on their crystal structure, electrical conductivity and thermal expansion.

EXPERIMENTAL

$(Mn,Co)_{3-x}Cr_xO_4$ (x=0, 0.5, 1, 1.5 and 2) spinel oxides were synthesized by solid state reaction. The mixtures of MnO (99%, Alfa Aesar), Co_3O_4 (99.83%, Fisher) and Cr_2O_3 (99%, Acros) were ball-milled for 48hr with de-ionized water, and dried overnight. The pre-mixed powders were pressed into pellets and bars, and then sintered in air at 1200°C for 24hr.

X-ray diffraction (XRD) was used for phase analysis by a Bruker D8 X-ray diffractometer. The samples were scanned in the 2θ range of 28-80° with the Cu K_α radiation at room temperature.

The electrical conductivity of spinel samples were measured by the four-probe dc method on rectangular bars with the dimension of about 14 x 6 x 2 mm^3 from 500 to 900°C in air. The thermal expansion coefficients (TEC) of the specimens were obtained using a Unitherm™ model 1161V high temperature vertical dilatometer from room temperature to 1000°C in air.

RESULTS AND DISCUSSION

XRD Phase Analysis

Figure 1 shows the XRD patterns of $(Mn,Co)_{3-x}Cr_xO_4$ (x=0, 0.5, 1, 1.5 and 2) spinel oxides at room temperature. $Mn_{1.5}Co_{1.5}O_4$ shows a mixture of cubic and tetragonal phases, which correspond to JCPDS No. 23-1237 and 18-0408. This result is consistent with Yang et al..[3, 4] As Cr content increased, the tetragonal phases decreased. When x reaches 1, $MnCoCrO_4$ showed almost cubic phase. When x is greater than 1, the cubic phase stabilized further. The structure was reported as No. 26-0474 in JCPDS.

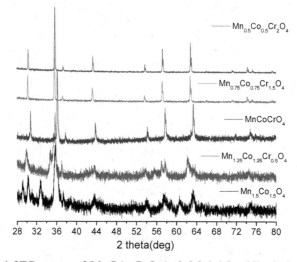

Figure 1. XRD patterns of $(Mn,Co)_{3-x}Cr_xO_4$ (x=0, 0.5, 1, 1.5 and 2) spinel oxides.

Electrical Conductivity Measurement

The effect of Cr on the conductivity of manganese cobalt spinel oxides at 500-900 °C in air is shown in Figure 2. As the Cr content increases, the conductivity decreases by several orders of magnitude. The conductivity increases with the temperature with a linear relationship between $\ln(cT)$ and 1000/T. This is consistent with the semiconductor behavior reported for other spinels.[3,] [7-9] and suggest that the conduction mechanism in the spinel oxides was the small polaron hopping mechanism in exchange pairs of Mn^{3+}/Mn^{4+} and Cr^{3+}/Cr^{4+} in octahedral sites.[8,10]

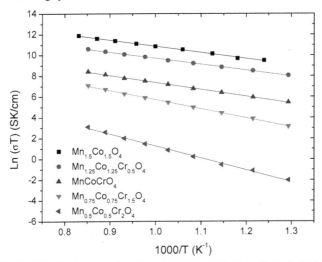

Figure 2. Electrical Conductivity as a function of 1000/T of $(Mn,Co)_{3-x}Cr_xO_4$ (x=0, 0.5, 1, 1.5 and 2) spinel oxides in air.

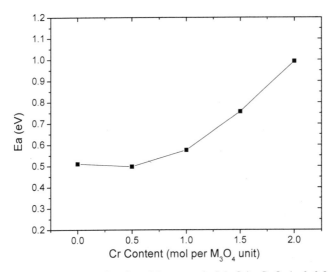

Figure 3. Activation Energy as a function of Cr content in $(Mn,Co)_{3-x}Cr_xO_4$ (x=0, 0.5, 1, 1.5 and 2) spinel oxides.

The small polaron hopping mechanism can be expressed as

$$\sigma = \frac{A}{T}\exp(-\frac{E_a}{k_b T})$$ (1)

Where, c is the electrical conductivity, T is the temperature, A is the pre-exponential factor, E_a is the activation energy, and k_b is the Boltzmann's constant. The activation energies obtained from the slope of the line in Figure 2 are shown as a function of Cr content in Figure 3. The activation energy increased with the Cr content for x greater than 0.05.

Thermal Expansion Test

Table 1 lists the TEC of $(Mn,Co)_{3-x}Cr_xO_4$ (x=0, 0.5, 1, 1.5 and 2) spinel oxides in air and shows that Cr doping decreases the TEC of Mn-Co-O spinel oxide. For comparison, the TEC of Cr_2O_3 from 25-900°C is 9.6 x10^{-6}/ °C.[2] Because the reaction layer exists between $Mn_{1.5}Co_{1.5}O_4$ coating and Cr_2O_3 scale outside the alloy interconnects after long-term operation, good TEC compatibility is required to reduce the thermal stress and avoid component spallation. The decrease in TEC of the Cr-containing reaction layer could lead to thermal stresses and local spallation.

Table I. TEC of $(Mn,Co)_{3-x}Cr_xO_4$ (x=0, 0.5, 1, 1.5 and 2) spinel oxides.

	TEC (20-1000°C, x10^{-6}/°C) in air
$Mn_{1.5}Co_{1.5}O_4$	10.8
$Mn_{1.25}Co_{1.25}Cr_{0.5}O_4$	9.94
$MnCoCrO_4$	8.80
$Mn_{0.75}Co_{0.75}Cr_{1.5}O_4$	7.51
$Mn_{0.4}Co_{0.6}Cr_2O_4$	6.76

CONCLUSION

$(Mn,Co)_{3-x}Cr_xO_4$ ($0 \leq x \leq 2$) spinel oxides were synthesized by solid state reaction. The effects of Cr doping on the crystal structure, electrical conductivity and thermal expansion were investigated. When Cr increased, the structure at room temperature changed from dual phase to cubic phase, the electrical conductivity decreased by several orders of magnitude, and TEC also decreased. The temperature dependence of the conductivity showed the semiconductor type behavior suggesting a small polaron hopping mechanism. These results suggest that the formation of a Cr-containing reaction layer between the oxidation scale and the spinel coating could lead to an increase in electrical resistance and thermal stresses. Further study is still needed to characterize the reaction layer.

ACKNOWLEDGEMENT

Financial support from the Department of Energy through the Building EPSCoR-State/National Laboratory Partnerships Program (Timothy Fitzsimmons, Program Office) is gratefully acknowledged.

REFERENCES
[1]S. C. Singhal, K. Kendall, High-temperature Solid Oxide Fuel Cells: Fundamentals, Design and Applications, *Elsevier Advanced Technology*, Oxford, 2003.
[2]N. Shaigan, W. Qu, D. G. Ivey, and W. Chen, A Review of Recent Progress in Coatings, Surface Modifications and Alloy Developments for Solid Oxide Fuel Cell Ferritic Stainless Steel Interconnects, *J. Power Sources*, **195**, 1529-42 (2010).
[3]Z. Yang, G. Xia, X. Li, and J. W. Stevenson, $(Mn,Co)_3O_4$ Spinel Coatings on Ferritic Stainless Steels for SOFC Interconnect Applications, *Inter J. Hydrogen Energy*, **32**, 3648-54(2007).
[4]Z. Yang, G. Xia, S. P. Simner, and J. W. Stevenson, Thermal Growth and Performance of Manganese Cobaltite Spinel Protection Layers on Ferritic Stainless Steel SOFC Interconnects, *J. Electrochem. Soc.*, **152**, A1896-1901 (2005).
[5]J. Fergus, K. Wang, and Y. Liu, Phase Equilibria in Spinel Coatings for Solid Oxide Fuel Cell Interconnects, pp. 679-687 in Materials Science & Technology 2009 Conference Proceedings, The Minerals, Metals and Materials Society, Pittsburgh, PA, 2009.
[6]J. Fergus, K. Wang, and Y. Liu, Interactions between $(Mn,Co)_3O_4$ SOFC Interconnect Coating Materials and Chromia, pp. 473-480 in 2010 The Minerals, Metals & Materials Society Annual Meeting Proceedings, The Minerals, Metals Materials Society, Seattle, WA, 2010.
[7]W. Qu, L. Jian, J. M. Hill, and D. G. Ivey, Electrical and Microstructural Characterization of Spinel Phases as Potential Coatings for SOFC Metallic Interconnects, *J. Power Sources*, **153**, 114-24 (2006).
[8]Z. Lu, J. Zhu, Electrical Conductivity of the Manganese Chromite Spinel Solid Solution, *J. Am. Ceram. Soc.*, **88**, 1050-3 (2005).
[9]R. Schmidt, A. Basu, and A. W. Brinkman, Small Polaron Hopping in Spinel Manganates, *Phys. Rev. B*, **72**, 1-9 (2005).
[10]N. M. Kovtun, V. K. Prokopenko, and A. A. Shamyakov, Electroconductivity and Electron Exchange in Spinel Structures, *Solid State Communications*, **26**, 877-8, (1978).

EFFECT OF METALLIC INTERCONNECT THICKNESS ON ITS LONG-TERM
PERFORMANCE IN SOFCS

Wenning Liu, Xin Sun, Liz Stephens, Moe Khaleel
Pacific Northwest National Laboratory
902 Battelle Blvd, Richland, WA. 99352

ABSTRACT
    At the operating environment of solid oxide fuel cells (SOFCs), oxide scale will
grow on the ferritic interconnect (IC) surface unavoidably and furfures induce growth
stress in oxide scale and along the interface of the oxide scale and IC substrate. A
combination of growth stress with thermal stresses may lead to scale
delamination/buckling and eventual spallation during SOFC stack cooling, even leading
to serious degradation of cell performance. In this paper, the effect of the ferritic IC
thickness on the delamination/spallation of the oxide scale was investigated numerically.
The predicted results show that the interfacial shear stresses increase with the growth of
the oxide scale and also with the thickness of the ferritic substrate; i.e., the thick ferritic
substrate can easily lead to scale delamination and spallation.

INTRODUCTION
    The operating temperature of solid-oxide fuel cells (SOFCs) has decreased from
1100°C to 800°C. This has led to a great deal of attention being focused on ferritic
stainless steel for use as an interconnect (IC) in SOFCs because of its gas-tightness, low
electrical resistivity, ease of fabrication, and cost-effectiveness [1]. In addition to
providing cell-to-cell electrical connections, ICs in SOFCs also act as separator plates in
separating the anode side fuel flow from the cathode side airflow for each cell. Compared
to chromium-based alloys, iron-based alloys have advantages in terms of high ductility,
good workability, and low cost. Iron-based alloys, especially Cr-Fe based alloys, e.g.,
Crofer 22 APU and SS 441, are by far the most attractive metallic IC materials for
SOFCs [2, 3]. However, under a typical SOFC working environment, an oxide scale will
grow on the metallic ICs in an oxidation environment, and furthermore, inducing growth
stress in the oxide scale and along the interface of the oxide scale and IC substrate. A
combination of the growth stresses with the thermal stresses induced by thermal
expansion coefficient mismatch between the oxide scale and the substrate may lead to
scale delamination/buckling and eventual spallation during stack cooling, which can lead
to serious cell performance degradation. Therefore, it is crucial to investigate the effect of
various geometric and mechanical factors on the reliability and durability of the metallic
interconnect in SOFC operating environments. As a powerful contender of ferritic
interconnects used in SOFC, the long-term performance of the oxide scale of SS 441 is
very important for its application in SOFCs.
    In this paper, the effect of the thickness of the ferritic substrate on the
delamination/spallation of the oxide scale on the ferritic substrate was investigated. The
cooling-induced stress in the oxide scale and on the interface of the oxide scale and the
ferritic ICs was predicted by using the finite element (FE) analysis. The predicted results
show that the interfacial shear stresses increase with the growth of the oxide scale and

also with the thickness of the ferritic substrate; i.e., the thick ferritic substrate can easily lead to scale delamination and spallation..

MATERIAL PROPERTIES AND FINITE ELEMENT MODEL

The SS 441 is one of the most powerful contenders for ferritic ICs used in SOFCs and is used in the present paper. The chemical composition is listed in Table 1 [4]. The surface roughness of the SS 441 sheet as-received was measured by profilometer. The surface roughness measurement is 0.39 μm with a standard deviation of 0.21 μm.

Table 1. Chemical composition of SS 441 (wt%)

Element	C	Mn	P	S	Si	Cr	Ni	Ti	Nb	N	Al	Fe
SS 441	0.009	0.35	0.023	0.002	0.34	18.0	0.3	0.22	0.5	0.014	0.05	rest

For uncoated SS 441 substrate, oxide scale will grow on the surface of the IC substrate and forms a bi-layer structure. Figure 1 depicts the bi-layer structures.

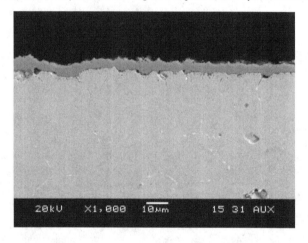

Figure 1. Bi-layer structure of oxide scale and SS 441 substrate.

During the cooling process from the 850°C in the furnace to the 25°C of room temperature, stress is induced in the oxide scale and the SS 441 substrate by the temperature drop because of the different CTEs of the oxide scale and the SS 441 substrate. FE simulations of the cooling process from the furnace to room temperature are performed to predict the cooling-induced stresses in the oxide scale and SS 441 substrate. It should be mentioned that in the simulation, only the cooling-induced stress was considered. The growth stress was not included. Therefore, the oxidization temperature is taken as stress-free temperature. The initial temperature is also taken as the oxidization temperature. Then the temperature of the oxidized SS 441 specimens is assumed to uniformly drop to the room temperature of 25°C. Taking advantage of the symmetrical nature of the sample, only a quarter specimen is considered here.

For the oxidized uncoated SS 441 substrate, the bi-layer system of the oxide scale and the SS 441 substrate was used. Figure 2 shows the cross-section of the typical model used for the bi-layer and tri-layer systems, respectively. For the bi-layer system of the oxide scale and SS 441 substrate, the thickness of the SS 441 substrate is taken as 0.25 mm, 0.50 mm, 1.0 mm, and 3.0 mm, respectively. The various thicknesses of the oxide scale were considered for each thickness of the SS 441 substrate.

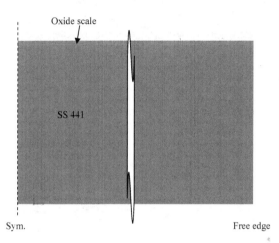

Figure 2. Cross-section of typical model: bi-layer system of oxide scale and SS 441 substrate

The commercial FE package, ABAQUS [5], was applied in simulating the static cooling process. An eight-node block solid element was used in this simulation. The typical FE mesh used for the SS 441/scale bi-layer system is illustrated in Figure 3. Very fine mesh was used near the interfaces between the substrate and oxide scale and also the sample edge to make certain that stress predictions were accurate at these locations.

The main composition of the oxide scale for SS 441 is Cr_2O_3, and the physical and mechanical properties for the oxide scale are taken as $\nu = 0.27$ and E=250 GPa [6, 7]. The mechanical properties of SS 441 are depicted in Figures 4 and 5 for temperature-dependent Young's modulus and CTE, respectively [8].

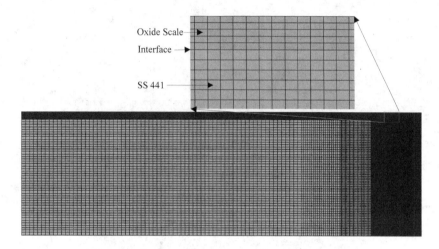

Figure 3 Typical mesh used in the modeling simulation

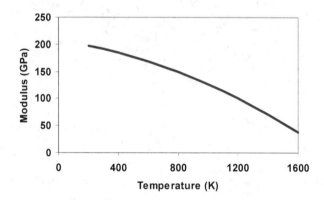

Figure 4. Temperature-dependent modulus of SS 441.

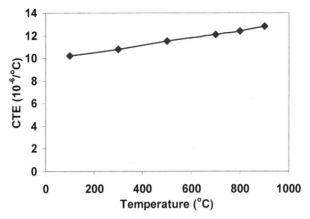

Figure 5. Temperature-dependent CTE of SS 441.

NUMERICAL RESULTS AND DISCUSSIONS

Oxidized, uncoated SS 441 was checked. Figure 6 shows the relationship between oxide scale thickness and the maximum interfacial shear stress for different substrate thicknesses. For a given substrate thickness, the shear stress increases with increasing oxide thickness, i.e., increasing the oxidation time, making cooling-induced delamination possible when the interfacial stress is greater than the interfacial shear strength. It is also interesting to note that for a given oxide thickness, i.e., oxidation time, the shear stress increases with increasing substrate thickness, indicating that the IC with a thicker substrate will have a greater tendency to delaminate and subsequently spall during cooling.

The cooling-induced compressive stress in the oxide scale versus the scale thickness for different substrate thicknesses is shown in Figure 7. For a given oxide thickness, i.e., a given oxidation time, the compressive stress in the scale increases with the substrate thickness. Since compressive stress in the scale is considered as the driving force for buckling-induced spallation during cooling, results here indicate that, at a given oxidation time, a thicker substrate has a higher tendency for spallation. It is also worth noting that, for all the cases considered, the scale compressive stress decreases with scale thickness, i.e., oxidation time. However, the overall level of the compressive stress in the scale remains high, ~2GPa.

The interfacial shear strength of the oxide scale and SS 441 substrate is determined as 391 MPa by an integrated experimental/modeling method. When the interfacial shear stress induced during the cooling down process exceeds this critical strength, delamination/spallation will occur. Figure 8 depicts a photo image of the oxidized, uncoated SS 441 upon removal from furnace after 600 h of oxidation at 850oC in air. The thickness of the SS 441 substrate is 1.5 mm, and the thickness of the oxide scale is measured as 2.8 μm by cross-section SEM. From Figure 6, it may be seen that the cooling induced interfacial stress is larger than the critical strength, therefore, the spallation of the oxide scale is unavoidable. The experimental observation validates the prediction of the numerical modeling.

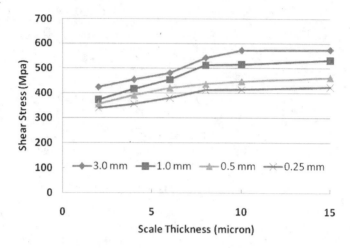

Figure 6 Cooling-induced interfacial shear stress versus oxide scale thickness for oxidized, uncoated SS 441.

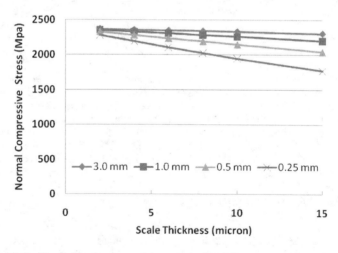

Figure 7 Cooling-induced compressive stress in the scale versus scale thickness for oxidized, uncoated SS 441.

Figure 8. Photo image of the oxidized, uncoated SS 441 upon removal from furnace after 600 h of oxidation at 850oC in air. Thickness of SS 441 is 1.5 mm.

CONCLUSIONS

The influence of the thickness of the SS 441 substrates on the interfacial shear stresses of the oxide/substrate interface, furthermore, the delamination/spallation of the oxide scale, was investigated by using finite-element. The uncoated SS 441 substrates with different thicknesses were studied numerically. The thickness of the oxide scale considered here varied from 2 μm to 15 μm in terms of different oxidization times. Based on the results of the modeling analyses, the following conclusions can be drawn: (1) the maximum interfacial shear stress between the oxide scale and the SS 441 substrate increases with the thicknesses of the oxide scale and the SS 441 substrate; (2) for a given substrate thickness, the shear stress increases with increasing oxide thickness, i.e., the increasing oxidation time, making cooling-induced delamination possible; for a given oxide thickness, i.e., oxidation time, the shear stress increases with increasing substrate thickness, indicating that an IC with a thicker substrate will have a greater tendency for delamination and subsequent spallation during cooling; (3) the normal compressive stress in the oxide scale increases with substrate thickness for a given oxide thickness, i.e., a given oxidation time; (4) for all the cases considered, the scale compressive stress decreases with scale thickness, i.e., oxidation time. However, the overall level of the compressive stress in the scale remains high, ~2GPa.

Though the thin SS 441 substrate leads to less tendency of delamination/spallation of the oxide scale, it has to be noted that the thin SS 441 substrate will result in greater warpage of the SS 441 substrate itself because of the growth of the oxide scale. Therefore, selection and design of the ferritic ICs in SOFCs should include the overall and comprehensive consideration of both delamination/spallation of the oxide scale and wargage of the ICs.

ACKNOWLEDGEMENTS
The Pacific Northwest National Laboratory is operated by Battelle Memorial Institute for the United States Department of Energy under Contract DE-AC06-76RL01830. The work was funded as part of the Solid-State Energy Conversion Alliance (SECA) Core Technology Program by the U.S. Department of Energy's National Energy Technology Laboratory (NETL).

REFERENCE
[1] J. W. Fergus, Metallic Interconnects for Solid Oxide Fuel Cells, Materials Science & Engineering A (Structural Materials: Properties, Microstructure and Processing) 397 (1-2) (2005) 271-83.
[2] L. Chen, E.Y. Sun, J. Yamanis, N. Magdefrau, Oxidation kinetics of $Mn_{1.5}Co_{1.5}O_4$ - coated Haynes 230 and Crofer 22 APU for solid oxide fuel cell interconnects, Journal of the Electrochemical Society, 157 (6)(2010) B931-B942.
[3] J. Rufner, P. Gannon, P. White, M. Deibert, S. Teintze, R. Smith, H. Chen, Oxidation behavior of stainless steel 430 and 441 at 800°C in single (air/air) and dual atmosphere (air/hydrogen) exposures, International Journal of Hydrogen Energy, 33(4)(2008) 1392-8.
[4] Product Data Bulletin, 441 Stainless Steel, UNS S44100 Preliminary Bulletin, AK Steel: http://www.aksteel.com/pdf/markets_products/Stainless/ferritic/441_Data_Bulletin.pdf
[5] ABAQUS/Standard User's Manual, Hibbitt, Karlsson and Sorensen Inc., 2002.
[6] A.M. Huntz, Stresses in NiO, Cr_2O_3 and Al_2O_3 oxide scales, Materials Science & Engineering, A201 (1995) 211-228.
[7] J.J. Barnes, J.G. Goedjen and D.A. Shores, Model for stress generation and relief in oxide-metal systems during a temperature change, Oxidation of Metals, 32(5/6) (1989) 449-469.
[8] Technical Data Blue Sheet, Stainless Steel AL 441HP Alloy, ATI Allegheny Ludlum: http://www.alleghenyludlum.com/ludlum/documents/441.pdf

CHARACTERIZATION OF THE CONDUCTIVE PROTECTION LAYERS ON ALLOY INTERCONNECT FOR SOFC

Xiaojia Du, Minfang Han[*], Ze Lei
Union Research Center of Fuel Cell, School of Chemical & Environment Engineering,
China University of Mining & Technology, Beijing, 100083, China

ABSTRACT

$La_{0.8}Sr_{0.2}Co_{0.6}Ni_{0.4}O_{3-\delta}$ (LSCN) powders were synthesized by the combustion method and characterized by physico-chemical analyses, such as X-Ray diffraction (XRD), field emission scanning electron microscope (FESEM), laser particle analysis and thermal dilatometry. LSCN protection layers were deposited on SUS430 and SUS410 alloys by screen printing technique and then sintered at different temperatures. The coated alloys were investigated in terms of high temperature oxidation behavior, conductivity and surface phase analysis. The results indicated that LSCN had a pure perovskite phase and uniform particle size distribution, as well as a good physical and chemical match with the electrolyte, cathode and interconnect materials. After 500h oxidation in air at 850°C, LSCN coating greatly improved the oxidation resistance of SUS430 and SUS410. The sample sintered at 950°C exhibited the lowest oxidation weight gain and the highest conductivity.

Keywords: SOFC; coating; LSCN; SUS430; SUS410

1. INTRODUCTION

Solid oxide fuel cells (SOFCs) have attracted a great deal of attention as clean power generation devices because of their high conversion efficiencies of fuel to electricity, low warm-gas emission and high system compactness [1-3]. The development of intermediate-temperature SOFC (IT-SOFC) can potentially reduce the manufacturing costs as it is operated at a lower temperature and may use inexpensive metallic interconnect [4]. Cr-based alloys, especially stainless steels such as the SUS430, are of low cost and have excellent machinability. However, the formation of chromia scales, especially in oxidizing atmospheres, leads to high contact resistance which is deleterious to the fuel cell performance. In addition, volatile Cr species can be released from the Cr_2O_3 scale, depending on the temperature and partial pressures of H_2O and O_2 [5]. The volatile chromium species, $CrO_2(OH)_2$, in the cathode of a SOFC will cause rapid poisoning of the cathode as well as the cathode/electrolyte interface, resulting in severe performance degradation[5-9]. To reduce the chromium vaporization under cathode operating conditions, either new oxidation resistant alloys with low chromium volatility or surface protective coatings on existing alloys are required [10-12]. The latter must possess a combination of physical and electrical properties to withstand the severe conditions under which SOFC is operated.

Conductive perovskites have been extensively investigated for coating material on metallic interconnects, such as $La_{0.85}Sr_{0.15}MnO_3$, $LaCrO_3$, $La_{0.8}Sr_{0.2}FeO_3$[13-15] etc. These materials have similar CTE to other components of SOFC and good ability to form a dense membrane. Besides, as a kind of rare earth element, La is useful in reducing the oxidation rate and improving the oxide-to-metal adhesion for chromia formers. $LaCoO_3$ has been found to exhibit a significant electronic conductivity but a higher CTE value than the other SOFC components. The substitution of cobalt with nickel was found to cause an increase in the electrical conductivity and a decrease in the thermal expansion [16], which may benefit the application of coating material. In this paper, $La_{0.8}Sr_{0.2}Co_{0.6}Ni_{0.4}O_{3-\delta}$ (LSCN) was selected as the coating material and investigated in terms of oxidation behavior, structural and electrical performance.

2. EXPERIMENTAL

Perovskite $La_{0.8}Sr_{0.2}Co_{0.6}Ni_{0.4}O_{3-\delta}$ (LSCN) powders were synthesized using the glycine-nitrate combustion method. Stoichiometric quantities of metal nitrates were dissolved in a minimum amount of distilled water. Glycine (total metal ion: glycine (mol ratio) =1:2) was added to the above thoroughly stirred solution. The mixed solution was heated in a big glass beaker at 80 °C to a viscous liquid, then heated to 200°C until spontaneous ignition of the liquid, forming a porous, foamy and fragile material. The precursor was calcined at different temperatures prior to characterization.

The ferritic stainless steel SUS 430, nominally 25×25×.5 mm, was used as the metallic interconnect material. LSCN coatings were fabricated on the modified steel surface by screen printing followed by heat-treatment at 850 ℃ in air. LSCN cylindrical samples, with diameter of 20mm and 5mm in length were used for CTE measurement. The oxidation behavior of SUS430 with and without coatings was evaluated via isothermal tests at 800 °C for 500 h in ambient air with 1-2% H_2O. Area-specific resistance (ASR) was measured using a four probe dc technique at 850 °C in air.

X-ray diffraction analyses on LSCN powders and oxidized alloys were carried out using a D/Max-RC X-ray generator with Cu Kα radiation. Scanning electron microscopy (SEM) analysis was conducted using a JEOL scanning electron microscope (model 6460) equipped with energy-dispersive spectroscopic capability at an operating voltage of 20 kV.

3. RESULTS AND DISSUSION
3.1 Powder Characterization

Fig. 1 shows the XRD patterns of the powders calcined respectively at 700 °C, 800 °C and 900 °C. The cubic perovskites could be obtained at all the three temperatures, for which there was little difference between the patterns. However, high temperature would cause increasing particle size, leading to poor sintering. Finally, 700 °C was selected as the sintering temperature. The particle size of LSCN, as seen in Fig. 2, was about 100-400 nm, which was consistent with the effective particle size, d_{50} of 313.0 nm measured by laser particle size analyzer (90 Plus, BROOKHAVEN, USA). The powders looked spherical and uniform in shape.

3.2 Effects of Chemical Compatibility of Coating

To examine its structure integrity under SOFC operation conditions, it is necessary to evaluate the chemical compatibility of coating with its adjacent stack components such as electrode and electrolyte. Fig. 3 and Fig. 4 show the XRD patterns of LSCN with YSZ and LSM at 850°C for 10h. As can be seen, no obvious reactions occurred between LSCN and YSZ or LSM, demonstrating good chemical compatibility of these materials.

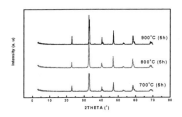

Figure 1. The XRD patterns of LSCN

Figure 2. Micrograph of LSCN

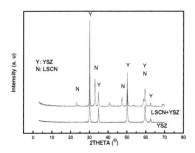

Figure 3. The XRD patterns of LSCN with YSZ at 850°C for 10h in air

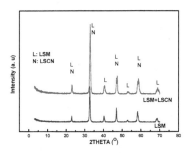

Figure 4. The XRD patterns of LSCN with LSM at 850°C for 10h in air

3.3 Thermal Expansion Measurement

To avoid spallation of coating from substrate, a good match of CTE between coating and alloy is needed. Fig. 5 illustrates the CTE curves of LSCN, SUS430, SUS410 and YSZ. The results of the differential thermal expansion measurements are listed in Table I. The CTE of LSCN is identical to those of both alloys, but a little higher than that of YSZ which may be due to the high CTE of element Ni.

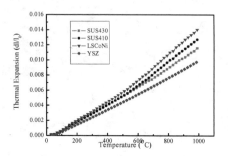

Figure 5. The CTE curves of LSCN ,YSZ, SUS430 and SUS410

Table I. The CTEs of LSCN ,YSZ, SUS430 and SUS410

Sample	SUS430	SUS410	LSCN	YSZ
CTE	11.3×10^{-6} K^{-1}	12.7×10^{-6} K^{-1}	13.3×10^{-6} K^{-1}	10.0×10^{-6} K^{-1}

3.4 Oxidation Kinetics

The coated coupons were sintered at 800°C, 950°C and 1150°C for 2h respectively, and then were isothermally oxidized at 850 °C for 500h. Fig. 6 and Fig. 7 show the oxidation weight gains of SUS430 and SUS410 with and without LSCN coating. According to Wagner's oxidation theory[17], the rate constant for oxidation, k_p, can be calculated as equation (1) and is listed in Table II.

$$\left(\frac{\Delta W}{A}\right)^2 = k_p \times t \qquad (1)$$

Where ΔW (mg) is the weight gain of sample at certain time , A is the surface area of sample (cm^2), k_p is the constant of oxidation rate (mg^2/cm^4·s), and t is the oxidation time (h).

As shown in Fig.6, the oxidation resistances of the alloys with time were greatly improved after the coatings were applied. The coated-coupons sintered at 950°C (LSCN-950) had the best results, with a k_p value of only 3.01✗0 $^{-13}$ mg^2/cm^4·s, and the weight gain was about 0.3mg/cm^2, approximately 1/5 of uncoated SUS430 sample. During long term oxidation, the grain size of LSCN-950 film would grow, which may be a mass barrier to block the Cr diffusion. But the coupons with LSCN coating sintered at 800°C had a higher k_p value of 6.08✗0 $^{-13}$ mg^2/cm^4·s, which may be attributed to the poor density of the coating layer, in which the porosity might not obstruct the inward oxygen movement completely; while the coupons with LSCN coating sintered at 1150°C, with a k_p value of 5.84✗0 $^{-13}$, were obviously over sintered. As shown in Fig. 7, the weight gain of coated SUS410 was similar to that of coated SUS430 in Fig. 6. The coated SUS410 sintered at 950°C showed the lowest weight gain, with a k_p value of 3.23✗0 $^{-13}$ mg^2/cm^4·s.

Fig. 8 and Fig. 9 show the XRD patterns of coated and uncoated coupons oxidized at 850°C for 500h. Compared with the uncoated sample, the coated samples formed less oxide which was mainly composed of Cr_2O_3 and $Mn_{1.5}Cr_{1.5}O_4$ on the surface of the alloys. This indicates the coatings inhibited the diffusion of Cr under operation conditions. To some extent small amount of Cr evaporated into the coating layer and reacted with Mn. After 500h isothermal oxidation at 850°C, the LSCN coating had good adherence to the alloys substrate, without any obvious spallation.

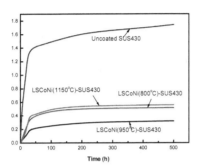

Figure 6. The weight gain of LSCN- SUS430

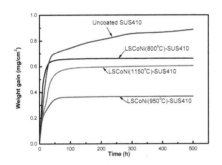

Figure 7. The weight gain of LSCN-SUS410

Table II. Oxidation rate constant $k_p(g^2/cm^4·s)$ for LSCN- SUS430 and LSCN-SUS410

SUS430	$k_p(g^2/cm^4·s)$	SUS410	$k_p(g^2/cm^4·s)$
LSCN(800)-SUS430	6.08×10 $^{-13}$	LSCN(800)-SUS410	5.65×10 $^{-13}$
LSCN(950)-SUS430	3.01×10 $^{-13}$	LSCN(950)-SUS410	3.23×10 $^{-13}$
LSCN(1150)-SUS430	5.84×10 $^{-13}$	LSCN(1150)-SUS410	4.67×10 $^{-13}$
Uncoated SUS430	2.42×10 $^{-12}$	Uncoated SUS410	6.11×10 $^{-13}$

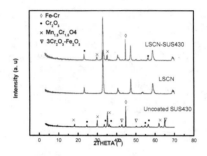

Figure 8. The XRD patterns of LSCN with SUS430 at 850°C for 500h

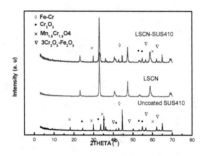

Figure 9. The XRD patterns of LSCN with SUS410 at 850°C for 500h

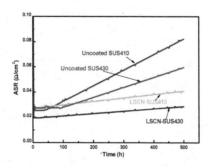

Figure 10. The ASR of coated and uncoated SUS430 and SUS410 at 850°C for 500h

3.5 Area Specific Resistance (ASR)

Fig. 10 presents the plot of ASR versus time at 850 $\mathbb{C}$ in air. As can be seen, the ASR values for both kinds of alloys had a considerable decrease after being coated with LSCN. The ASR of bare coupons increased rapidly as a function of time, while the coated samples had a much smaller increase in ASR, especially for the coated SUS430 coupon which showed a low and relative stable ASR of $20 m\Omega\cdot cm^2$ after 500h, implying better performance of oxidation resistance than that of the coated SUS410. However, both coupons demonstrated a trend of increase in ASR after the oxidation; this may be attributed to the oxidation of the alloys as well as the Cr diffusion from interconnect. Further study will be done to figure out the densification of the coatings and long term stability of coated alloys.

4. CONCLUSIONS

$La_{0.8}Sr_{0.2}Co_{0.6}Ni_{0.4}O_{3-\delta}$ (LSCN) was successfully synthesized and prepared on ferritic stainless steels SUS430 and SUS410 by screen printing method. It demonstrated good chemical compatibility with adjacent components and good thermal expansion match to the alloys. After 500h aging in air, the ASR of the coated coupons was significantly lower than that of the bare alloy. LSCN coated SUS430 samples had the lowest ASR of 20 $m\Omega\cdot cm^2$. It is concluded that LSCN can be a promising candidate material for the protective coating on metallic interconnect in IT-SOFC.

ACKNOWLEDGEMENTS

Financial support from the NSFC key project (50730004) of China and MOST project (2009 DFA 6136) of China are appreciated. It is appreciated to Dr. S. C. Singhal for reviewing the paper.

FOOTNOTES

Corresponding author: hanminfang@sina.com

REFERENCES

[1] Minh, N.Q., Ceramic Fuel Cells, Journal of the American Ceramic Society, 76(3), 563-588(1993).

[2] Singhal, P. and N.Q. Minh, Solid oxide fuel cells: Technology status, International Journal of Applied Ceramic Technology, 1(1), 5-15(2004).

[3] Singhal, S.C., Advances in solid oxide fuel cell technology, Solid State Ionics, 135(1-4), 305-313(2000).

[4] Jong Seol Yoon, Jun Lee, Hae Jin Hwang et al, Lanthanum oxide-coated stainless steel for bipolar plates in solid oxide fuel cells (SOFC), Journal of Power Sources, 181, 281-286(2008).

[5] Asteman, H., et al., Indication of chromium oxide hydroxide evaporation during oxidation of 304L at 873 K in the presence of 10% water vapor, Oxidation of Metals, 52(1-2), 95-111(1999).

[6] Badwal, S.P.S. et al., Interaction between chromia forming alloy interconnects and air electrode of solid oxide fuel cells, Solid State Ionics, 99(3-4), 297-310(1997).

[7] Jiang, S.P., J.P. Zhang, X.G. Zheng, A comparative investigation of chromium deposition at air electrodes of solid oxide fuel cells, Journal of the European Ceramic Society, 22(3), 361-373(2002).

[8] Matsuzaki, Y. and I. Yasuda, Electrochemical properties of a SOFC cathode in contact with a chromium-containing alloy separator, Solid State Ionics, 132(3-4), 271-278(2000).

[9] Matsuzaki, Y. and I. Yasuda, Dependence of SOFC cathode degradation by chromium-containing

alloy on compositions of electrodes and electrolytes, Journal of the Electrochemical Society, **148**(2), A126-A131(2001).

[10]Yang, Z.G., J.W. Stevenson, and K.D. Meinhardt, Chemical interactions of barium-calcium-aluminosilicate-based sealing glasses with oxidation resistant alloys, Solid State Ionics, **160**(3-4), 213-225(2003).

[11]Yang, Z.G., G.G. Xia, and J.W. Stevenson, $Mn_{1.5}Co_{1.5}O_4$ spinel protection layers on ferritic stainless steels for SOFC interconnect applications, Electrochemical and Solid State Letters, **8**(3), A168-A170(2005).

[12]Stanislowski, M., et al., Reduction of chromium vaporization from SOFC interconnectors by highly effective coatings, Journal of Power Sources, **164**(2), 578-589(2007).

[13]Jong-Hee Kim, RaK-Hyun Song, Sang-Hoon Hyun. Effect of slurry-coated $LaSrMnO_3$ on the electrical property of Fe-Cr alloy for metallic interconnect of SOFC, Solid State Ionics. **174**, 185-191(2006).

[14] Jong Seol Yoon, Jun Lee, Hae Jin Hwang at el. Lanthanum oxide-coated stainless steel for bipolar plates in solid oxide fuel cells (SOFC), Journal of Power Sources, **181**, 281-286(2008).

[15]X. Montero, et al, Comparative study of perovskites as cathode contact materials between an $La_{0.8}Sr_{0.2}FeO_3$ cathode and a Crofer22APU interconnect in solid oxide fuel cells, Journal of Power Sources, **188**, 148–155(2009).

[16]V.V.Kharton, A.V. Kovalevsky, V.N. Tikhonovich at. el, Mixed electronic and ionic conductivity of $LaCo(M)O_3$ (M=Ga, Cr, Fe, or Ni), Solid State Ionics, **110**, 53-60(1998).

[17] Morin, F., G. Beranger, and P. Lacombe, Limits of Application for Wagners Oxidation Theory, Oxidation of Metals, **4**(1), 51(1972).

Novel Cell/Stack
Design and Processing

ADVANCED MANUFACTURING TECHNOLOGY FOR SOLID OXIDE FUEL CELLS

Norbert H. Menzler, Wolfgang Schafbauer, Robert Mücke, Ralf Kauert, Oliver Büchler, Hans Peter Buchkremer and Detlev Stöver

Forschungszentrum Jülich
Institute of Energy and Climate Research, IEK-1
52425 Jülich, GERMANY

ABSTRACT

SOFCs lack entry to the market, firstly due to the manufacturing costs of the components and secondly due to high degradation rates.

FZ Jülich has been working on SOFCs for approx. 20 years. The focus of the last years was the development of cells with high power density, reduced degradation of SOFC stacks, and the introduction of industrial-applicable manufacturing technologies.

Typical anode-supported cells (ASCs) were manufactured by tape casting the support and screen printing of the functional layers. At IEK-1 a new manufacturing technology was introduced. The complete half-cell was tape cast, but in reverse order. This means that firstly the electrolyte is cast, then the anode is cast on the electrolyte, dried and afterwards the substrate is cast on the double layer system. The half-cell is then cut in green state and sintered. By adapting carefully the powders used for the functional layers and the substrate also the flattening procedure can be omitted. Thus, the complete cell is producible by casting the electrolyte, the anode and the substrate on each other, cutting, sintering to electrolyte gas-tightness and afterwards cathode coating by screen printing followed by a final sintering step. Manufacturing of such cells for single cell and stack testing revealed appropriate electrochemical results Due to relatively thick electrolyte and anode (~ 20µm each) the power densities are reduced by approx. 10-20% in comparison to state-of-the-art ASCs.

1   INTRODUCTION

Market entry of SOFC systems is up to now hindered by two hurdles: firstly the manufacturing costs of all components (cells, stack, BoP) and secondly the still too short lifetimes (or too high degradation rates). Especially for stationary applications a lifetime of more than 40,000 h of operation is aimed for and thus degradation rates, in terms of voltage loss, should be less than 0.5%/1000 h. Therefore much R&D work focuses on reducing the manufacturing costs and enhancing lifetime of SOFC components. Main focal points are:

Higher power density (higher power density reduces the amount of components needed and thus reduces costs per system);

Reduced component sizes to minimize materials costs (mainly reduced cell and interconnect thickness);

Reduction of high-temperature processes during cell manufacturing;

Implementation of mass production techniques for the cell and the metal parts

Reduced system operating temperature to minimize thermally activated degradation phenomena and to introduce cheaper metallic materials.

All of the above mentioned goals must be reached by keeping the basic functionality and requirements of the cell, the stack and the whole system. Basic requirements are:

Cell   -   Electrochemical, physical and chemical functionality
        -    Manufacturability (materials interactions, component integrity)

	-	Power output
	-	Costs
Stack	-	Functionality (electrochemical, contacting)
	-	Gas-tightness
	-	Cyclability
	-	Long-term durability
	-	Costs
System	-	Functionality (efficiency, fuel and air utilization)

- Component adjustment (according to temperature, temperature differences, cyclability, pressure differences, gas stream distribution, process measuring and control technology)
- Thermocyclability
- Costs

This paper deals with the core component, the cell, and the possibility to reduce manufacturing costs while maintaining high power densities at the same time. The basic manufacturing techniques are explained, the wet chemical route presented in deeper manner and subsequently demonstrated how modern industrial-applicable production technologies and additional supporting scientific R&D can reduce the manufacturing costs.

2   MANUFACTURING TECHNOLOGIES

The manufacturing technologies are divided into techniques for substrate and for cell fabrication. A more deeper insight can be taken in[1], here a short overview will be given. In figure 1 an overview about possible manufacturing technologies for SOFC substrates and functional layers is presented. The substrate (irrespective of anode, electrolyte, cathode or metal-supported) is manufactured mostly by powder technological techniques like pressing or casting (tape, slip). The layers were applied by either thermal techniques like plasma spraying (atmospheric, vacuum, low pressure) or high velocity oxy fuel spraying, or more classical ceramic technologies like screen printing, casting or powder spraying. Additionally, mostly for special layers, thin film techniques were also used (e.g. physical or chemical vapour deposition, sputtering, electrophoresis). In recent years also the sol-gel technique was applied successfully. Which technology will be used at least for each substrate type or functional layer depends mostly on the envisaged thickness, microstructure, roughness and so on. Each technology has its own advantages and limitations:

Thermal technologies:

Are established in various R&D fields (e.g. thermal barrier coatings for turbine blades)
Can apply layers either on metallic or on ceramic supports
Typically the layer thickness is too high (SOFC functional layers mostly require less than 20 μm)
The layer microstructure is not ideal; for the electrodes the structure is too dense and with mostly splat like pores parallel to the layer and for the electrolyte less gas tight (in comparison to layers applied by ceramic technologies and sintering)

Applied R&D with thermal technologies are in the field of SOFC the coating of metallic interconnects with protection, contacting or electrically insulating layers and for the coating of metal-supported cells with all functional layers.

Wet chemical (ceramic) technologies:

are established in many application fields (ceramics for electronics, electrics, bio medicine...)

a broad variety of different technologies is available and can be combined

to obtain mechanical strength, materials stability, adhesion, conductivity and microstructure a high-temperature sintering step is needed

a cutting step is typically needed for shaping (laser cutting, water jet, punching); this can be done partly in green state (if near-net-shape manufacturing is possible) but partly also in end-fired state

The wet chemical (ceramic) route of cell manufacturing is nowadays established in many fields of application. This route is used on one hand for the planar or tubular supports and on the other hand for the functional layers. Additionally also protection and contact layers on interconnects were applied by such techniques.

Thin film technologies

are established in special application fields (CDs, DVDs, semi conductors)

have minimal materials costs due to thin structures, but frequently high invest costs due to expensive equipment

often require a special microstructure and surface which can be coated (e.g. for very thin and gas-tight layers a smooth substrate with no or very small pores is essential)

Thin film technologies were used today in the SOFC field for example for interdiffusion barrier or electrically insulating layers (between 8YSZ and LSCF cathode or on interconnects).

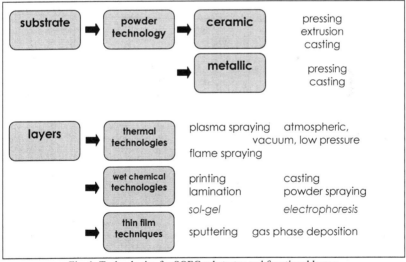

Fig. 1: Technologies for SOFC substrates and functional layers

3 MANUFACTURING OF ANODE-SUPPORTED SOFCS

In figure 2 a typical manufacturing line for anode-supported large-scale (200x200mm²) SOFCs at Forschungszentrum Jülich used until 2005 is presented.

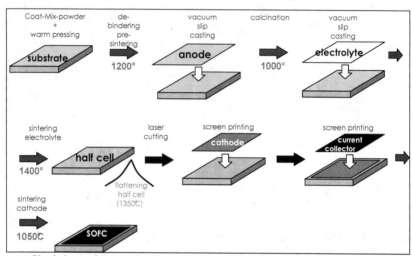

Fig. 2: State-of-the-art of cell manufacturing at Forschungszentrum Jülich in 2005

The manufacturing in 2005 can be characterized by a) using a broad variety of techniques (most of them discontinuous) and b) in summary four sintering steps at high temperatures (above 1000°C) to ensure high quality cells with high power density. The high amount of high-temperature sintering steps and the goal for mass applicable manufacturing techniques were the motivation to optimize the cell manufacturing including cost reduction. Based on this, the goals for the development between 2005 and 2010 were:

  reduced materials inventory (e.g. thinner anode substrate, in 2005 1500µm)
  reduced number of different manufacturing technologies needed for a cell
  implementation of techniques with high throughput
  reduction of the number of sintering steps
  reduction of maximum sintering temperature needed
  introduction of cheaper materials

And all this by maintaining
  the excellent current density, and
  the manufacturability into the common stack design

The main focal points were executed in parallel and were:
  understanding of the sintering behaviour of thin, dense layers (electrolyte) on thick, porous support; these investigations were used to understand exactly the sintering process for evaluating potentials to better adapt the sintering characteristics of support and layers. Additionally, it should be investigated whether the flattening step can be omitted by carefully adapting the sintering regime[2, 3]

implementation of industrial compatible manufacturing techniques like tape casting and roller coating[4-6]

green-in-green manufacturing to reduce number of sintering steps[5]

4   SINTERING OF MULTI-LAYERED STRUCTURES

Figure 3 presents the bending of a three layer structure composed of anode substrate, anode and electrolyte during sintering, cooling and subsequent NiO reduction. It can be seen that the laminate tends firstly to bend concave (due to higher sintering activity of the electrolyte material). At a given time after the maximum sintering temperature of 1400°C has been reached, the electrolyte becomes dense and finishes to shrink, however, the substrate is still porous and shrinks further. Due to this ongoing shrinkage the multi-layered structure now bends convex. During cooling down and the subsequent NiO reduction this convex-type bending is enhanced (CTE of 8YSZ is lower than of 8YSZ/NiO cermet). During heat-up for operation the convex bending is diminished but not minimized to zero.

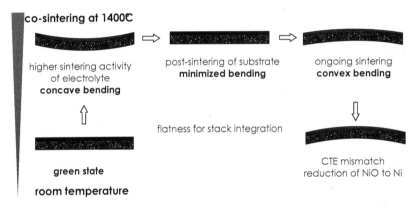

Fig. 3: Bending behavior of substrate-anode-electrolyte layer system during thermal and chemical processes[7]

The sintering behaviour of the single half-cell components was subsequently characterized more in detail[2, 3]. It is obvious (Fig. 4) that irrespective of the pre-treatment of the electrolyte material its sintering behaviour cannot be adapted completely to that of the substrate. A sintering mismatch of approximately 100-200K remains.

The experiments highlighted in Fig. 4 were carried out by using free-standing layers (substrate 1-1.5 mm, electrolyte 300 μm thick). Also by varying the sintering regimes a sintering mismatch between the support and the layer remains.

Additionally to the experiments the remaining sintering bending was modelled, see Fig. 5. The modelling presents beginning at around 1000°C, a concave bending of the structures due to higher electrolyte sintering activity.

The bending is largest at the beginning of the highest sintering temperature of 1400°C and reduces during the holding time at T_{max}. After approx. 1-2 h of holding time the bending reaches zero; during the additional dwell time the component starts to bend convex due to post-sintering of the substrate and reaches values of ~ 5m[-1]. During the sintering the substrate possesses a linear shrinkage of ~12% while the substrate reaches 17%. As the coefficient of thermal expansion of the substrate is larger than the one of the electrolyte, the 8YSZ layer exhibits a compressive residual stress after manufacturing. This is a main advantage of such processing route because typically the cells exhibit a quite low bending strength (50-150 MPa). But during cell assembling into the stack the cells must withstand handling and flattening. Therefore, a "supporting" compressive stress in the electrolyte ensures better cell integrity. The compressive stresses reach values of up to 500 MPa[8].

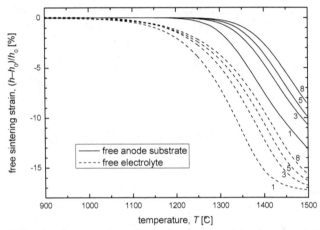

Fig. 4: Relative sintering expansion of the anode substrate and the electrolyte during sintering; the numbers denote the heating rates in K/min

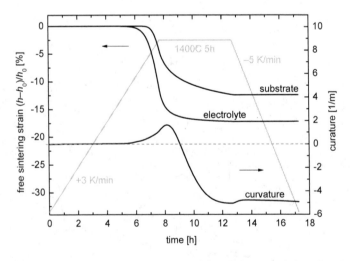

Fig. 5: Simulation of the bending behaviour of anode substrate and electrolyte during sintering

In an additional test series the influence of the substrate on the densification of the electrolyte was examined. For that, sintering experiments were carried out on free-standing electrolyte layers and on electrolyte layers on rigid and on shrinking substrates (dense and porous). It was found out that 8YSZ can not be densified to more than 90% (of theoretical density) on rigid substrates. Free-standing layers can be densified at 1400°C, but layers on co-shrinking substrates reach a higher density at the same sintering conditions than the free-standing single layers. The fact that a layer, sintered on a co-shrinking support, densifies higher than a free-standing material was confirmed with these experiments the first time. This result implies that a sufficient densification of the 8YSZ substrate at 1400°C is only obtainable by sintering it on a parallel shrinking support (Fig. 6). If the basic 8YSZ material (chemically synthesized) was replaced by a sintered and crushed material, it could be sufficiently densified at 1400°C.

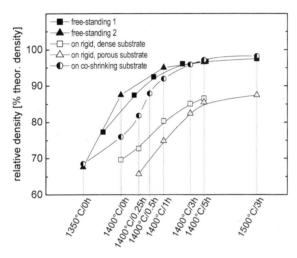

Fig. 6: Relative density of 8YSZ electrolyte layers in various configurations[3] (log θ, which is the x-axis is an effective sintering work based on the master sintering curve model[9])

5   IMPLEMENTATION OF ROLL COATING TECHNOLOGY

Up to now mostly slip casting and screen printing have been used for SOFC functional layer coating. Especially the discontinuous (vacuum) slip casting process is a pure lab-scale process and can not be adapted to industrial throughput. But it was used because of its ideal applicability to various layer thicknesses and film microstructures. To verify whether industrial-applicable coating technologies can be applied to SOFC manufacturing the roll coating process was evaluated[4]. For the coating a so called reverse roll coater from Mathis (model: RRC BW, Mathis AG, Switzerland) was used. In this process rotating rolls transport the substrate and apply the slurry on the substrate (Fig. 7). Either green or pre-sintered substrates can be used for anode and electrolyte coating.

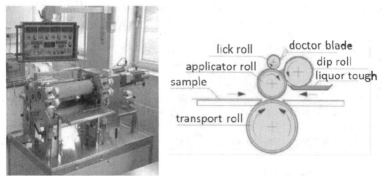

Fig. 7: Roll coater (RC), left: general view, right: scheme of RC process

In figure 8 the microstructure of an electrolyte on a substrate (green-in-green) applied by roll coating is presented. It is obvious that a relative high amount of pores remains after sintering. But He-leak testing reveals sufficient gas tightness of the layer, which implies that the visible porosity is mostly closed. Additionally it can be concluded that the applied electrolyte is still too thick (aimed thickness 5-10µm, actual 15µm)

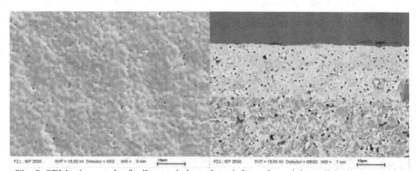

Fig. 8: SEM micrograph of roll coated electrolyte; left: surface, right: polished cross section

Cells with a roller coated electrolyte exhibited in single cell test (50x50mm² samples) a current density of approx. 1 A/cm² (at 800°C and 0.7V with hydrogen/air). A standard anode-supported cell from FZ Jülich yields 1.4 A/cm² at comparable conditions. The reduced current density can be explained with a higher ohmic resistance due to the thick electrolyte and with a non-optimized anode microstructure. The optimization of the process (layer thickness and microstructure) is still in progress.

6   GREEN-IN-GREEN TAPE CASTING OF HALF-CELLS

Tape casting is established in SOFC manufacturing as technology to produce the support[1, 10]. Within a series of experiments aiming for better adapted tape-casting slurry, a complete novel half-cell (substrate + anode + electrolyte) manufacturing technology could be established[5]. Traditionally the functional layers are coated on top of the tape-cast substrate. Thereby, the surface structure (porosity, roughness, waviness) governs the thickness and microstructure (density/porosity) of the next layer. With the new manufacturing process the processing sequence was changed. Starting with tape casting of the thin electrolyte the surface of this layer becomes nearly ideal because it is cast on a very smooth, defined polymeric support tape. After the electrolyte dried, the anode is tape cast on top of the electrolyte and subsequently after a next drying step the thick anode substrate is cast on the double layer. The main advantage of this "reverse coating" is that now the functional layers electrolyte and anode can be cast in its envisaged thickness and homogeneity. Especially thickness variations of the substrate which are typically in the range of 3-5% of the overall thickness (at 300 μm thickness this means a thickness change of up to 15 μm) can be higher than the applied layer thickness (e.g. 10 μm). In Fig. 9 this geometrical problem is explained. The half-cells cast by this method had an overall thickness of about 600μm. In our technical scale manufacturing, the green-on-green casting was realized by rolling the dried layer back on the primary transport roll and moving it afterwards again under the doctor blade (and again the bi-layer electrolyte and anode). In an industrial tape casting bench this can be realized by three doctor blades, one after each other, and in between short drying chambers (~ 1 m).

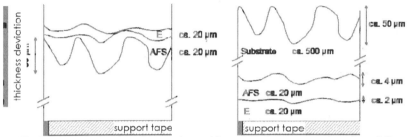

Fig. 9: Schematic drawing of the coating problem of thin layers on rough surfaces; left: traditional manufacturing route, right: novel reverse route (E: electrolyte, AFS: anode)

Due to the equipment the anode and the electrolyte could not be cast as thin as intended. The intended thickness for the anode was 7 μm and for the electrolyte 10 μm; the actual thicknesses were ~ 20 μm. For the electrochemical characteristics this results in lower power density due to higher ohmic resistances. By adapting the cast head (special doctor blades) or introducing a micro tape caster this problem can be overcome. In Fig. 10 the casting results from a substrate cast directly on an electrolyte and a polished cross section of such bi-layer is presented.

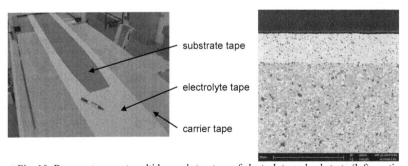

Fig. 10: Reverse tape-cast multi layered structure of electrolyte and substrate (left: casting result in green state, right: SEM polished cross section)

The laminates composed of electrolyte, anode, and substrate were cut or punched in green state after drying. Afterwards samples of dimensions of 100 x 100 mm² were sintered at normal conditions (5 h at 1400°C). After sintering the samples were characterized with respect to their bending by using a laser based optical inspection system[11, 12] (Fig. 11a). The samples bent drastically (6.5 mm) and showed some "Sombrero-like" warpage. This special type of bending is caused by the fact that, at maximum temperature, the gravitation force is opposite to the bending direction and the materials became soft and subsequently the inner part of the half-cell inclines. This behaviour is the same for the traditional manufacturing route (but not that pronounced) and is overcome by "ironing" the half-cell at high temperatures by putting a dead load on the samples. By varying the starting electrolyte powder (by replacing the chemically synthesized material by a sintered and crushed powder) we successfully reduced the bending during sintering without adding any additional flattening step (Fig. 11 b)[13]. The bending is now less pronounced and more uniform (formed like a cylinder section). Cells with such bending can be mounted without problems into the state-of-the-art Jülich stationary stack design.

Through variation of manufacturing sequence, manufacturing technology and materials we successfully omitted the pre-sintering step for the anode substrate, the calcination of the anode and the flattening step after electrolyte sintering. The amount of sintering steps during cell manufacturing is reduced from five to two (half cell and cathode sintering).

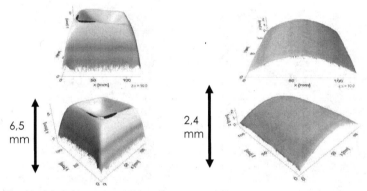

Fig. 11: a) left: laser topographic measurement of the bending of an SOFC half cell after final sintering; b) right: variant with alternative electrolyte material

With this type of manufacturing technique multi-layered structures of electrolyte and substrate as well as multi-layered structures of electrolyte, anode and substrate were manufactured. Single cells based on such half-cells exhibited a current density of approx. 1 A/cm² (800°C, 0,7V, H_2 / air). The lower current density is again due to the higher ohmic resistance because if the thicker functional layers. Cells with a size of 100 x 100 mm² were manufactured and assembled into a Jülich F-design stack successfully. The electrochemical characteristics of this short-stack were similar to comparable stacks with cells manufactured by the traditional way.

In figure 13 the novel manufacturing route is presented finally.

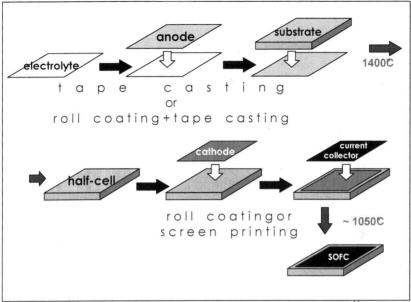

Fig. 13: Novel manufacturing route for anode-supported SOFCs 2010[14]

7   SUMMARY AND OUTLOOK

With intensive R&D in the fields of understanding sintering of layered laminates composed of porous and dense ceramic structures, adaption of processing technologies and sequences and the basic materials characteristics we successfully produced planar anode-supported SOFCs. The novel route is simplified and cheaper than the traditional one. It is based on only two manufacturing technologies, tape casting and roll coating, and two sintering steps at high temperatures. This route has the potential to fulfil the requirements stated out by industry for SOFC market entry.

Future work will focus on the reduction of the overall cell thickness (goal: 250-300 μm), enhancing the reproducibility of the manufacturing steps, the scale-up to cell sizes for power stacks (e.g. 200x200 mm²) and the tape casting of the electrolyte and the anode with a special thin film tape casting bench.

REFERENCES

[1]Menzler N.H., Tietz F., Uhlenbruck S., Buchkremer H.P., Stöver D., J. Mat. Sci. **45** Iss. 12 (2010), 3109-3135

[2]Mücke R., PhD thesis Ruhr-University Bochum 2007. Writings of Forschungszentrum Jülich, Series Energy & Environment, Part 9, ISBN 978-3-89336-529-6 (2008), available in German only

[3]Mücke R., Menzler N.H., Buchkremer H.P., Stöver D., J. Am. Ceram. Soc. **92** (2009), S95-S102

[4]Büchler O., Bram M., Mücke R., Buchkremer H.P., Electrochem. Soc. Trans. 25 (2) (2009), 655-663

[5]Schafbauer W., PhD thesis Ruhr-University Bochum 2009. Writings of Forschungszentrum Jülich, Series Energy & Environment, Part 66, ISBN 978-3-89336-631-6 (2010)

[6]Büchler O., Schafbauer W., Mücke R., Menzler N.H., Proc. 9th European Fuel Cell Forum, June 29 – July 02 2010, Lucerne, Switzerland, 17-67 – 17-73 (2010)

[7]Steinbrech R.W., Malzbender J., Buchkremer H.P., Menzler N.H., Blum L., de Haart B.,

[8]Groß S.M., Steinberger-Wilckens R., CD-ROM Proc. of the 6th Int. Symp. SOFC at the 33nd Int. Conf. & Exp. Adv. Ceram. Comp. (2009), Daytona Beach, USA

[9]Malzbender J., Fischer W., Steinbrech R.W., J. Power Sources 182 (2008), 594-598

Su H., Johnson L, J. Amer. Ceram. Soc. 79 No. 12 (1996), 3211-3217

[10]Menzler N.H., Zahid M., Buchkremer H.P., In Kriegesmann J. (Ed.): Technical Ceramic Materials. Publisher Deutscher Wirtschaftsdienst. Chapter **3.4.6.1** (2004), 1-21, available in German only

[11]Mücke R., Menzler N.H., Buchkremer H.P., Stöver D., Mat. Sci. Forum Vols. **654-656** (2010), 2410-2413

[12]Mücke R., Menzler N.H., Kemnitzer F., Blöchl K., Proc. 8th European Fuel Cell Forum, June 30 – July 04 2008, Lucerne, Switzerland

[13]German Patent Application DE 102009040297.7-45

[14]Menzler N.H., Büchler O., Mücke R., Schafbauer W., Buchkremer H.P., Stöver D., In Kriegesmann J. (Ed.): Technical Ceramic Materials. Publisher Deutscher Wirtschaftsdienst (2010), accepted for publication, available in German only

PRODUCTION OF CURRENT COLLECTOR-SUPPORTED MICRO-TUBULAR SOLID OXIDE FUEL CELLS WITH SACRIFICIAL INNER CORE

Ricardo De la Torre, Michele Casarin and Vincenzo M. Sglavo
Department of Materials Engineering and Industrial Technologies, University of Trento, 38123 Trento, Italy

ABSTRACT

A novel current collection method for micro-tubular solid oxide fuel cells (SOFCs) was studied in the present work. Commercial pencil leads were used as supports for the current-collector during processing and as sacrificial inner core for the fabrication of micro-tubular cells. In the present innovative current-collection configuration a nickel wire was coiled around commercial pencil lead thus producing a compliant and homogeneously distributed current collector along the cell length and with two terminals available for the electrical connections. The cell consists of NiO/YSZ anode, YSZ electrolyte and LSM cathode sequentially deposited by dip coating. After removing pyrolizable organics in air, the nickel wire was protected from oxidation under argon atmosphere upon sintering. I-V curves showed maximum power output of $\sim$190 mW/cm^2 at 800°C under wet H$_2$ fuel flow in cells with $\sim$1.2 mm diameter and 30 mm length with an active cathode area of about 0.75 cm^2.

INTRODUCTION

Solid oxide fuel cells (SOFCs) are energy conversion devices that produce electricity from chemical reactions between a fuel and an oxidant across an ionic conducting oxide electrolyte at high temperatures with high conversion efficiency, fuel flexibility and low emissions.[1-3] Materials such as Ni-YSZ (yttria stabilized zirconia) cermet for the anode, YSZ for the electrolyte and lanthanum strontium manganite (LSM) for the cathode are commonly used in SOFCs production.[4,5] Since the operating temperatures of SOFCs are relatively high (800-1000°C), many efforts have been focused on lowering them through the implementation of new configurations, materials and fuels. SOFCs may be fabricated in a variety of design configurations and geometries but the planar and tubular geometric configurations are the most recognized. Tubular cells offer many advantages over planar cells such as rapid start-up/shut-down time, higher thermal shock resistance, easy sealing and excellent thermo-cycling stability.[6,7] Moreover, decreasing the tubular cell diameter in the millimeter or sub-millimeter range makes would allow to achieve substantial increase in volumetric power density of the cell stacks.[8,9] Nevertheless, the much longer current path in a tubular than in planar SOFC represents one of the most important issues due to its significant effect in the ohmic losses especially for the electrode support (usually the anode), where the electrical connection is often produced via wires placed within the tube or wounded on areas of the anode not covered by the electrolyte. Therefore, the current collector-method in tubular cells is absolutely crucial for the efficiency of the cell. Some studies have proved that an increase of the surface of the current collector in contact with the electrode leads to much reduced ohmic losses and current paths.[10,11]

The substrate/support of the tubular SOFCs is often fabricated by traditional extrusion techniques,[12] but other techniques are also applied such as slip casting[13] and dip coating.[14] Compared with extrusion, slip casting and dip coating techniques have many advantages including easy processing, low cost and possibility to fabricate complex shapes. Commercial pencil leads have been used for many purposes showing satisfactory results not only as pyrolizable material but as electrode or matrix because it is easily machined and offer good mechanical properties at high temperatures.[15-17]

In the present investigation, we report the conditions to produce current collector-supported micro-tubular SOFCs of about 1.2 mm in diameter by dip coating by using commercial pencil lead used as sacrificial inner core.

EXPERIMENTAL PROCEDURE

The current-collector cell support was fabricated by coiling a nickel wire with 0.1 and 0.05 mm diameter (99.994%, Alfa Aesar GmbH & Co KG, Karlsruhe, Germany) around a 0.5 mm diameter pencil lead, 50 mm long (Faber Castell, Super polymer, 9065 S-2B, Germany).

NiO (J.T. Baker Inc., USA) and 8 mol% Y_2O_3-ZrO_2 (TZ-8YS, Tosoh, Japan) powders were used as raw materials for the production of anode and electrolyte. Ammonium polyacrylate (Darvan 821A, R.T. Vanderbilt Company Inc., USA) was used as dispersant in the slurry. High molecular weight acrylic based polymers in aqueous emulsion (Duramax B-1000 and B-1014, Rohm and Haas Co., France) were used as binder.

NiO/YSZ (58/42 weight ratio) powders were dispersed in distilled water and mixed for 2 h in a high energy rotatory mill (Turbula T2F, Bachofen, Switzerland). In order to increase the porosity of the anode layer, graphite powder (Flake, 7-10 μm, Alfa Aesar, Germany) was added (from 5 to 10 wt%) to the NiO/YSZ slurry. The binder was added during a gently magnetic stirring after milling. Solid load content in the anode slurry was varied between 60 and 90 wt%.

The nickel current-collector cell support was dipped into the anode slurry and then dried at 100°C for 1 h. The process was repeated until the Ni wire was completely covered. The porosity of the anode was determined by the Archimedes's method.

The same dip coating and drying conditions were used to deposit the electrolyte onto the anode though, in this case, only one immersion was required. The thickness of the electrolyte layer was controlled through the dwelling time and the solid content of the slurry.

The produced half-cells were kept at 800°C for 1 h in air to remove the organic additives, graphite pore former and the inner core (pencil lead); then a flux of Ar was injected into the furnace chamber to complete the co-sintering at 1380°C for 2 h.

Suspensions of $La_{0.8}Sr_{0.2}MnO_{3-\delta}$ (LSM) (Fuel Cell Materials, USA) and LSM/YSZ (weight ratio of 1:1) in α-Terpineol were used as conductive and functional layer for the cathode, respectively. Sintered half-cells were coated with the functional layer and subsequently with the conductive layer by dip coating. Every layer was followed by drying at room temperature for about 12 h. The bilayer-structured cathode was sintered at 1150°C for 2 h in Ar after the elimination of organics at 800°C in air.

The size of the produced cells was ~1.2 mm diameter and ~30 mm length with active cathode length of ~20 mm, whose active cell area was 0.75 cm^2.

Scanning electron microscopy (SEM) (JEOL, JSM 5500, Japan) was used to observe the microstructure of the micro-tubular cells.

The single cell performance was investigated using a potentiostat (Autolab PGSTAT302N, Eco Chemie, Netherlands). The electrochemical measurement was carried out in the temperature range from 700 to 800°C. Both edges of the cell were sealed to alumina tubes by a ceramic sealant (Ultratemp 516, Aremco, Valley Cottage, NY). For current collection on the anode side both nickel wire ends already exposed outside the cell were attached to platinum wires (99.997%, Alfa Aesar, Germany) using silver paste. A platinum gauze (99.9%, 0.1 mm diameter wire, 25x25 mm, Alfa Aesar, Germany) placed on whole cathode area was used to collect the current on the cathode side. The electrical contact on the cathode surface was increased by silver ink.

Hydrogen (humidified by bubbling water at room temperature) was flowed within the tubular cell at 25 mL/min. The cathode side was open to the air without flowing gas. The current-voltage (I-V) curves were recorded by linear sweep voltammetry at scanning rate of 5 mV s^{-1}.

RESULTS AND DISCUSSION

Figure 1 shows the current collector-supports produced with two different wire diameters. In both cases the current-collector support shows about 10-11 turns per centimeter after coiling.

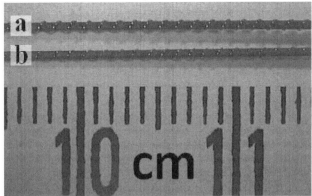

Figure 1. Current-collector cell supports with different wire diameters: (a) 0.1 mm; (b) 0.05 mm

The mechanical resistance of the two considered wires was checked after a heat treatment at 800°C for 1 h under air atmosphere. In both cases the wires were barely oxidized after the heat treatment but only the thicker wire maintained sufficient mechanical resistance for successive processing and handling.

Current-collector supports were dipped into anode slurries with different solid loading content. As the dip/pull out speed was kept constant, the number of dips was only dependent of the thickness obtained in each immersion and on the slurry solid content. Figure 2 shows optical photographs of the anode layer deposited onto the current-collector support from slurries with different solid loading. The number of immersions was determined as a function of the minimal thickness required to get a uniform diameter along the Ni support. The anode layer deposited with solid contents of 60 and 70 wt% shows several cracks due to many dip and drying steps. On the other hand, with a solid content of 90 wt% only 3 dips were necessary to obtain an adequate thickness, though a quite large deposit in each dip generated bubbles and radial cracks upon drying. A crack-free anode layer was obtained with 80 and 85 wt% solid content after 4 and 3 dips, respectively; therefore, solid content of 85 wt% was fixed to fabricate the cells with minimum number of dips.

The effect of graphite as pore former within the anode was evaluated after fixing the solid loading in the NiO-YSZ slurry. Figure 3 shows the apparent porosity produced in the anode after sintering with different amount of graphite. The porosity in the anode supports (open porosity) must be enough high to facilitate the flux of gas species through the layer, but not too much to reduce the triple phase boundary (TPB) at the interface anode-electrolyte.[18,19] Therefore, 33.5% porosity achieved with 5% of graphite was considered the most appropriate to fulfill such requirements.

Similarly to the anode, an analysis of solid loading content in YSZ slurries was performed. Here, slurries with 60, 70 and 80 wt% were prepared and deposited onto the anode by dip coating. As shown in Figure 4, the thickness of the electrolyte increases with higher solid loading. The electrolyte layer deposited from the slurry at 60 wt% (Fig. 4 (a)) separated after the sintering from the anode due to a poor adhesion at the interface. Conversely, the adhesion of the layers with solid loadings of 70 and 80 wt% (Fig. 4 (b) and (c), respectively) is clearly superior; moreover, the electrolyte layer at 70 wt% is thinner as well, with clear benefits for the cell performance.

The remains of the pencils leads were analyzed on the half-cells after the sintering. The main drawback of using pencil leads as fugitive material is the presence of by-products that can cause a reduction in the anode porosity. Figure 5 shows that these impurities (wax and resins) are not easily detected on the microstructure. Furthermore, the majority of impurities are deposited only in small zones lengthwise of the cell and always on the inner surface of the anode. Therefore, merely a slight effect on the performance of the cell should be expected.

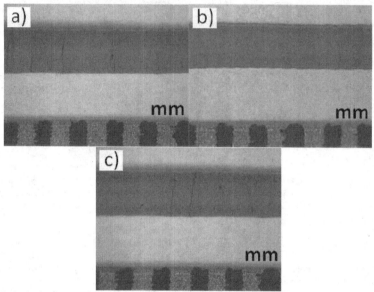

Figure 2. Optical micrographs of anode layer deposited onto current-collector supports from slurries at different solid loading content: (a) 60-70 wt%; (b) 80-85 wt%; (c) 90 wt%.

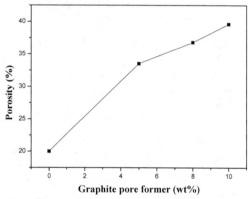

Figure 3. Effect of graphite pore former on the porosity of anode layer sintered at 1380°C.

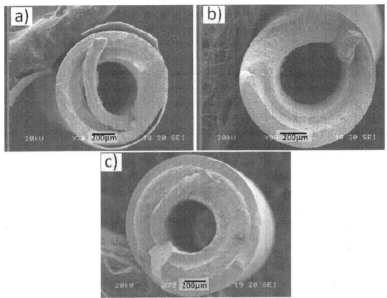

Figure 4. Scanning electron microscopic micrographs of cross-section view of half-cells sintered at 1380°C. The electrolyte was deposited from slurries with different solid loading: (a) 60 wt%; (b) 70 wt%; (c) 80 wt%.

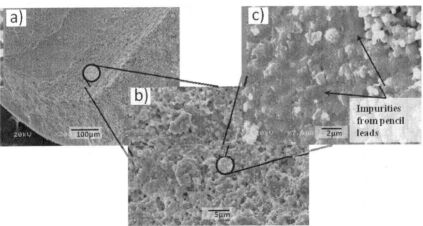

Figure 5. Scanning electron microscopic micrographs of the remains deposited onto the inner anode surface: (a) longitudinal-section view; (b), (c) surface view at different magnifications.

In order to complete the cells fabrication, LSM/YSZ and LSM cathode inks were applied onto the sintered half-cells by dip coating. Figure 6 shows the complete cell after cathode sintering at 1150°C. The availability of the two exposed ends from the embedded current collector to be connected for the electrochemical testing is clear. The maximum diameter of the cell can oscillate between 1.1 and 1.2 mm.

Figure 7 shows an overview of the current collector-supported micro-tubular cell microstructure. The thickness of the anode and cathode is about 200 μm and 45 μm, respectively. The electrolyte with a thickness of about 15 μm is shown.

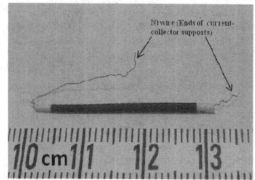

Figure 6. Photograph of current collector-supported micro-tubular cell.

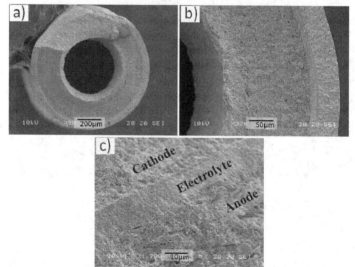

Figure 7. Scanning electron microscopic micrographs of a complete cell: (a) overview; (b) cross-section view; (c) larger magnification for cross-section view.

The electrochemical performance of the current collector-supported micro-tubular cell is shown in Figure 8. Maximum peak power densities of 56, 112 and 190 mW/cm^2 were obtained at 700, 750 and 800°C respectively. The open circuit voltage was stable at 0.93V.

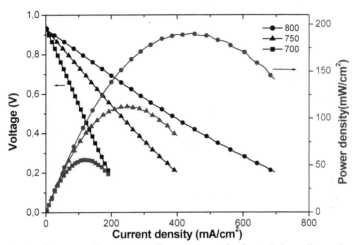

Figure 8. Cell performance of a current collector-supported micro-tubular cell tested at different temperatures.

CONCLUSIONS

In this study, current collector-supported micro-tubular cells with diameter about 1.2 mm were successfully produced and characterized. This novel current-collection method makes possible a reduction of the ohmic resistance as a result of shorter current paths within the inner electrode of the cell. Moreover, the two exposed wires outside the cell offer an easy way to connect these small devices with the terminals of the set-up equipment.

REFERENCES
[1]N. Q. Minh, Solid Oxide Fuel Cell Technology-Features and Applications, *Solid State Ionics*, **174**, 271-77 (2004).
[2]R. N. Singh, Sealing Technology for Solid Oxide Fuel Cells (SOFC), *Int. J. Appl. Ceram. Technol.*, **4** [2] 134-44 (2007).
[3]A. Dhir and K. Kendall, Microtubular SOFC Anode Optimisation for Direct Use on Methane, *J. Power Sources*, **181**, 297-303(2008).
[4]S. Giraud and J. Canel, Young's Modulus of Some SOFCs Materials as a Function of Temperature, *J. Eur. Ceram. Soc.*, **28**, 77-83 (2008).
[5]M. Mogensen and S. Skaarup, Kinetic and Geometric Aspects of Solid Oxide Fuel Cell Electrodes, *Solid State Ionics*, **86-88**, 1151-60 (1996).
[6]L. Zhang, H. Q. He, W. R. Kwek, J. Ma, E. H. Tang and S. P. Jiang, Fabrication and Characterization of Anode-Supported Tubular Solid Oxide Fuel Cells by Slip Casting and Dip Coating Techniques, *J. Am. Ceram. Soc.*, **92** [2] 302-10 (2009).
[7]J. Ding and J. Liu, Fabrication and Electrochemical Performance of Anode-Supported Solid Oxide Fuel Cells by a Single-Step Cosintering Process, *J. Am. Ceram. Soc.*, **91** [10] 3303-07 (2008).

[8]T. Suzuki, T. Yamaguchi, Y. Fujishiro and M. Awano, Improvement of SOFC Performance Using a Microtubular, Anode-Supported SOFC, *J. Electrochem. Soc.*, **153** [5] A925-28 (2006).

[9]T. Suzuki, Y. Funahashi, T. Yamaguchi, Y. Fujishiro and M. Awano, Fabrication and Characterization of Micro Tubular SOFCs for Advanced Ceramic Reactors, *J. Alloys and Compd.*, **451**, 632-35 (2008).

[10]D. Cui, L. Liu, Y. Dong and M. Cheng, Comparison of Different Current Collecting Modes of Anode Supported Micro-Tubular SOFC through Mathematical Modeling, *J. Power Sources*, **174**, 246-54 (2007).

[11]T. Suzuki, T. Yamaguchi, Y. Fujishiro and M. Awano, Current Collecting Efficiency of Micro Tubular SOFCs, *J. Power Sources*, **163**, 737-42 (2007).

[12]C. Yang, W. Li, S. Zhang, L. Bi, R. Peng, C. Chen and W. Liu, Fabrication and Characterization of an Anode-Supported Hollow Fiber SOFC, *J. Power Sources*, **187**, 90-92 (2009).

[13]J. Ding, J. Liu, W. Yuan and Y. Zhang, Slip Casting Combined with Colloidal Spray Coating in Fabrication of tubular Anode-Supported Solid Oxide Fuel Cells, *J. Eur. Ceram. Soc.*, **28**, 3113-17 (2008).

[14]J. C. Ruiz-Morales, J. Peña-Martínez, J. Canales-Velázquez, D. Marrero-López, C. Savaniu and P. Núñez, Cost-Effective Microstructural Engineering, *J. Am. Ceram. Soc.*, **92** [1] 276-79 (2009).

[15]M. Perdicakis, H. Aubriet and A. Walcarius, Use of a Commercially Available Wood-Free Resin Pencil as Convenient Electrode for the 'Voltammetry of Microparticles' Technique, *Electroanalysis*, **16** [24] 2045-50 (2004).

[16]D. F. Rohlfing and A. Kuhn, Scanning Tunneling Microscopy of Electrode Surface Using Carbon Composite Tips, *Electroanalysis*, **19** [2-3] 121-28 (2007).

[17]C. Black, C. Poile, J. Langley and J. Herniman, The Use of Pencil Lead as a Matrix and Calibrant for Matrix-Assisted Laser Desorption/Ionisation, *Rapid Commun. Mass Spectrom.*, **20**, 1053-60 (2006).

[18]S.P. Jiang, P.J. Callus and S.P.S. Badwal, Fabrication and Performance of Ni/3mol% Y2O3-ZrO2 Cermet Anodes for Solid Oxide Fuel Cells, *Solid State Ionics*, **132**, 1-14 (2000).

[19]D. Dong, M. Liu, Y. Dong, B. Lin, J. Yang and G. Meng, Improvement of the performances of tubular solid oxide fuel cells by optimizing co-sintering temperature of the NiO/YSZ anode-YSZ electrolyte double layers, *J. Power Sources*, **171**, 495-98 (2007).

Reliability/Degradation

NUMERICAL MODELING OF CATHODE CONTACT MATERIAL DENSIFICATION

Brian J. Koeppel, Wenning Liu, Elizabeth V. Stephens, and Moe A. Khaleel
Pacific Northwest National Laboratory
Richland, WA, USA

ABSTRACT
Numerical modeling was used to simulate the constrained sintering process of the cathode contact layer during assembly of solid oxide fuel cells (SOFCs). A finite element model based on the continuum theory for sintering of porous bodies was developed and used to investigate candidate low-temperature cathode contact materials. Constitutive parameters for various contact materials under development were estimated from dilatometry screening tests, and the influence of processing time, processing temperature, initial grain size, and applied compressive stress on the free sintering response was predicted for selected candidate materials. The densification behavior and generated stresses within a 5-cell planar SOFC stack during sintering, high temperature operation, and room temperature shutdown were predicted. Insufficient constrained densification was observed in the stack at the proposed heat treatment, but beneficial effects of reduced grain size, compressive stack preload, and reduced thermal expansion coefficient on the contact layer densification and stresses were observed.

INTRODUCTION

The joining of SOFC stack components using contact materials is critical to ensure that electrical and mechanical integrity is maintained during operation. Planar SOFC stacks require reliable uniform contact to carry the electrical current between the series-connected cells for good performance. On the cathode side, the contact layer must survive the oxidizing environment to maintain a durable bond between the ceramic electrode and the metallic interconnect. Stable but expensive noble metals have been used for contact materials, but less expensive alternatives are being investigated including perovskites with additions to aid sintering[1-3]. Here the contact layer is often completed during multiple stack assembly steps, so processing temperatures are limited for this interface to avoid excessive oxide scale growth when a metallic interconnect is used. Ceramic contact materials are desirable for their compatibility with the cathode and good oxidation resistance, but it is difficult to form a dense mechanical bond at temperatures below the typical range for good sintering. Also, densification of the contact layer during assembly will result in volumetric changes that create residual stresses in the stack. Densification and other complex material behaviors during stack fabrication have not typically been considered in thermal-stress analyses of SOFCs but do create an initial stress state. An understanding of these densification effects on the load distribution and stresses in the stack during assembly is necessary to quantify the mechanical reliability of the contact layer. Furthermore, if the structural integrity of the contact layer can be improved sufficiently to carry and distribute the thermal mismatch loads between the cell and interconnect, mechanical reliability of the entire cell and seal layers can possibly be improved. For example, load transmission from the cell to the ductile metallic interconnect was demonstrated to beneficially reduce the transmitted load through the cell's rigid perimeter seal[4].

The objective of this work was to develop a numerical model for sintering of contact materials integral to the reliable assembly of SOFC substrates and stacks. The model accounts for the effect of kinematic constraints and remote loading on the local densification behavior. The model also predicts the generated residual stresses in the contact layer due to densification which occur in addition to the typical thermal-mechanical stresses induced at the operating temperature. The results of these modeling analyses will help stack designers reduce high stresses in the stack so that structural failures are prevented and high stack mechanical reliability is achieved to meet performance targets.

NUMERICAL MODEL DESCRIPTION

The technical approach taken for this study was to develop the necessary numerical modeling procedures for simulating densification strains and transfer them to PNNL's existing SOFC modeling tools. The inclusion of densification strains with the thermal strains resulting from the SOFC operating temperature profile would provide a more accurate assessment of the stack's stress state.

Constitutive Model

Potential constitutive models were researched in the literature and evaluated for implementation to capture the mechanical effects of contact paste densification. The continuum viscous sintering model[5] was selected as suitable for simulating volumetric changes of the material by densification in the three-dimensional stack modeling framework. This phenomenological constitutive model is a general nonlinear-viscous continuum model based on plastic deformation of porous materials that computes the inelastic strain rates as a function of grain size, relative density, mechanical properties, stress, and temperature. The linear viscous form of the model is given in Eq. (1) where c_{ij} is the stress, $\dot{\varepsilon}_{ij}$ is the strain rate, η_0 is the shear viscosity of the fully dense material, and δ_{ij} is the Kronecker delta. The terms ψ and φ are the normalized bulk and shear viscosities, respectively, with the assumed dependence on porosity θ given by Eq. (2) and Eq. (3). The term P_L is the 'sintering stress' given by Eq. (4) where a is the surface energy and r is the average grain radius. This continuum-level model can capture the enhanced sintering behavior that occurs under applied stress and the influence of the mechanical constraint from the material layers being joined. For the special case of free sintering, the relation reduces to that shown in Eq. (5).

$$\sigma_{ij} = 2\eta_0\left[\varphi\dot{\varepsilon}_{ij} + \left(\psi - \frac{\varphi}{3}\right)\dot{\varepsilon}_{kk}\delta_{ij}\right] + P_L\delta_{ij} \tag{1}$$

$$\psi = \frac{2}{3}\frac{(1-\theta)^3}{\theta} \tag{2}$$

$$\varphi = (1-\theta)^2 \tag{3}$$

$$P_L = \frac{3a}{r}(1-\theta)^2 \tag{4}$$

$$\eta_0 = -\frac{P_L}{6\psi\dot{\varepsilon}_{11}} \tag{5}$$

This model was then implemented into the MSC MARC general purpose finite element analysis (FEA) software[6] through user-defined subroutines. The numerical model response was tested by evaluating the influence of material, initial grain size, processing temperature, processing time, and applied load on the densification rate of simple specimen geometries.

Stack Geometry Model

Next, the contact layer densification model was implemented in a thermal-structural analysis for a five cell planar co-flow stack. The stack consisted of five 25×25 cm^2 active area repeat units using ribbed interconnect plates and was constrained between thick upper and lower load plates. The densification model was applied to a 400 m contact layer between the cathode and the interconnect ribs, where the contact paste was assumed to be LSCF plus 3mol% CuO with an initial grain size $2r$ of 1.0 m and relative density of 51%. A uniform compressive pressure of 0.2 MPa was applied to the top surface of the upper load plate in the stack. Due to symmetry, only half of the stack was modeled. The densification of the layer was simulated at a 900°C fixed temperature heat treatment for two hours. The stress state induced by the densification was computed, followed by predictions of the stack stress states at the operating condition and during the shutdown process to room temperature. The operating temperature profile (Figure 1) was computed independently and corresponds to a stack operating at ~400 mA/cm^2 using 97% H$_2$ fuel, 80% fuel utilization, and 12% air utilization within a 700°C furnace operating condition. The global minimum and maximum temperatures in the solid components at the operating condition were 703°C and 823°C, respectively (Figure 1).

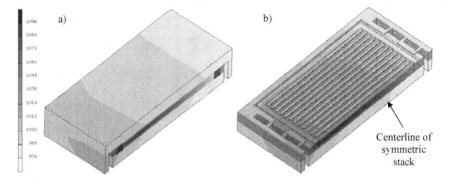

Figure 1. Operating temperature profile (K) for the 5-cell co-flow generic stack shown a) with and b) without the top plate. Tmin=703°C (976°K) and Tmax=823°C (1096°K).

MATERIAL PROPERTIES DEFINITION

PNNL's ongoing development activity of in-house contact materials[1] used dilatometric testing under a constant heating rate to screen candidate materials. This simple test was used to estimate key parameters for the densification model. For example, the strain as a function of temperature for a candidate material (LSM plus 3mol% CuO+BaCuO$_2$ as a sintering aid) is shown in Figure 2a. Assuming that the strain is attributed entirely to densification under no load (i.e. free sintering), the densification strain rate as function of temperature is computed. Based on the aforementioned constitutive model, this strain rate is directly related to the shear viscosity of the fully dense material η_0 as shown in Eq. (5). Particle growth was not experimentally measured but was expected to be small for low processing temperatures. Experimental data for LSM from the literature[7] confirmed low growth rates for the temperature range of interest. The initial grain size of 0.35 m was estimated by microscopy and a growth rate expression based on the literature data was implemented in the model. The surface energy α for such ceramics depends on orientation and is difficult to measure, but the values for ceramics and metals are generally near unity[8] so a constant value of 1.2 J/m^2 was assumed. The shear viscosity parameter η_0 was then calculated as a function of temperature (Figure 2b). No densification occurs at low temperatures, so the viscosity term is capped for numerical stability. This procedure was also applied to other candidate materials such as LNF and LSCF with various sintering aids. The material parameters were then implemented in the FEA constitutive model and the original response curve from the experiment was verified (Figure 2a). This single test does not determine all of the required properties to fully characterize the material for simulation, but this available data is presently adequate for modeling until the candidate materials are down-selected and more thoroughly tested.

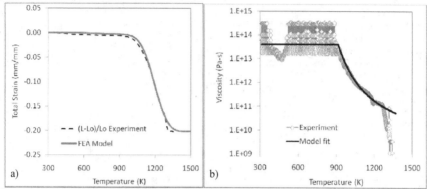

Figure 2. Experimental data and fitted model results for a) the linear densification strain and b) the temperature dependent viscosity of LSM plus 3mol% CuO+BaCuO$_2$.

MODELING RESULTS AND DISCUSSION

First, the densification behavior of selected candidate contact materials currently under development were evaluated to determine the predicted influence of initial grain size, processing temperature, processing time, and applied load on the densification rate. Higher densities are desirable since the bulk strength of porous materials increases with relative density.

Free Sintering Response of Selected Contact Materials

Using the material data obtained from the dilatometry tests, the densification behaviors of the LSM-based, LNF-based, and LSCF-based contact materials were compared under free sintering conditions (Figure 3). It was assumed that the processing temperature must be limited with metallic interconnects to no more than 1000°C for durations up to four hours. At the upper end of this temperature range, the LSM-based material was predicted to reach nearly full density. The LSM-based material showed a strong temperature sensitivity indicating that the sintering aid had reduced the acceptable temperature range significantly to achieve strong sintering rates around 950°C. It was also observed that the predicted grain growth of the LSM-based material was small from an initial 0.35 m average size to 0.44 m, which is consistent with the general observations of no apparent grain growth in the experiments. Also at the upper temperature range, the LSCF-based material was still more than 90% dense, but the LNF-based material was less than 81% dense. At the lower end of this temperature range, which is likely closer to the temperatures to be realistically used with current cell and seal technologies, the LSCF-based material showed the best response followed by the LSM-based and LNF-based materials. Even with processing at 800°C, the LSCF-based material still increased its relative density to 58-65%. This suggests that further sintering may even be possible during stack operation for cell regions where local temperatures exceed 800°C.

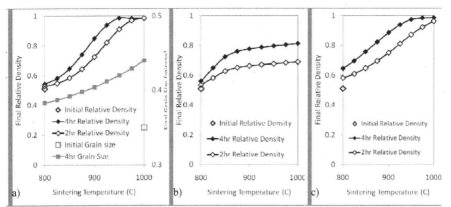

Figure 3. Predicted final density for a) LSM-based, b) LNF-based, and c) LSCF-based contact materials using two hour and four hour heat treatments with processing temperatures of 800-1000°C.

Effect of Initial Grain Size on Free Sintering Response

The effect of initial grain size on the final density was evaluated for the LSM-based material at a 950°C processing temperature (Figure 4). Smaller grain sizes increase the inherent sintering stress as indicated in Eq. (4). A strong sensitivity to achieve nearly full density was observed for grain sizes less than 0.5 m. These small grained materials exhibited rapid sintering without excessive grain growth to achieve high density. If precursor materials can be economically milled to these sizes and actual grain growth rates are low as assumed, then this approach may be advantageous to achieve suitable low temperature densification.

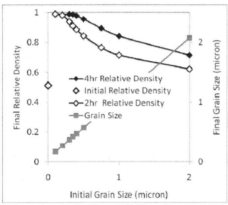

Figure 4. Final relative density and grain size for an LSM-based contact material as a function of initial grain size (950°C processing temperature).

Effect of Compressive Preload on Sintering Response

Remote compressive stresses can provide additional driving force to aid the inherent sintering stress and increase the densification rate. The effect of remote uniaxial stress was evaluated for the LSM-based material with a two hour 950°C heat treatment (Figure 5). With compressive stress on the specimen, the sintering strain rates in the axial and lateral directions now differ. As the axial stress increases, more axial densification strain occurs while the lateral strain decreases and may even become positive with large applied stresses. The impact on the densification rate is small though for low applied stresses. The final density for the two hour treatment with 1-2 MPa applied stress increased to only 92-93% relative density compared to 91% for free sintering. Applied stresses less than the sintering stress of ~5 MPa showed only small improvements, while stresses near or greater than the sintering stress were much more beneficial. However, imposition of such high stresses uniformly across the contact region of actual large area cells is likely impractical for tall stacks though.

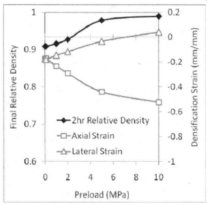

Figure 5. Final relative density and densification strains for an LSM-based contact material as a function of uniaxial preload (950°C processing temperature)

Densification in Stack Structures

The effect of kinematic constraints is important for sintering of SOFC layers in the stack. For materials sintered to other substrates, the bond with the other surface restricts the in-plane sintering strain while amplifying the out-of-plane sintering strain. This results in variation of not only the sintering strain rates but also the local stresses that they depend upon and ultimately the final local density. Therefore, constrained sintering layers are expected to exhibit spatial variation in density and residual stresses depending on the local properties and geometry of the contact and substrate materials.

The final relative density for the LSCF-based contact material is shown in Figure 6, which indicates that the constraints of the bonded substrates have significantly hindered the densification. The final relative density varied from 51-55% indicating little to no densification as compared to the comparable free sintering condition where a final density of 65% was predicted. The densification was also highly non-uniform. The region near the center of the stack showed the highest densification while the corners and edges showed less densification. The center of the stack is more flexible to accommodate the shrinkage of the contact layer compared to the edges where out-of-plane displacement is restricted by the stiff support frame around the cell. Tensile stresses are then developed here during the heat treatment which counter the inherent sintering stress and reduce the densification rate. It was also observed that the density distribution showed little variation between the cells.

Figure 6. Contours of the final relative density for the LSCF-based cathode contact layers in the planar cell stack geometry.

Previous in-house mechanical testing of low temperature contact materials[9] had suggested maximum tensile strengths of the order of 15 MPa; however, predicted stresses on the contact layers were observed to locally exceed this value. The maximum principal stresses for the contact layers are shown in Figure 7, where only tensile stresses in excess of the estimated 15 MPa limit are indicated for clarity. These regions are the areas that would likely physically separate during sintering to leave a void or mechanically fail by cracking during operation and shutdown. At the end of the sintering heat treatment, the contact layer remains in compression for most of the layers except the ends of the ribs for the upper cells (Figure 7a). With the addition of thermal strains due to the operating temperature profile, these regions of high stress at the end of the ribs enlarge for all of the cell layers (Figure 7b). The stack could still operate electrochemically with material failure in these regions, but the local loss of electrical contact may result in reduced performance and higher current densities within the cell. The shutdown condition predicts large regions of high stress (Figure 7c). This occurs in the region that was hottest during operation and experiences the largest thermal strains during shutdown. Such large regions of contact failure/interface delamination would not be acceptable for tolerance to planned or unplanned thermal cycling of the stack. Therefore, the model captures the contact layer stresses during the densification process, though the example generic stack geometry used here would not satisfactorily tolerate expected operational procedures.

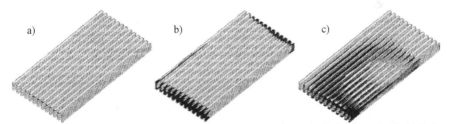

a)   b)   c)

Figure 7. Contours of the maximum principal stress in excess of 15 MPa limit (shown in black) during a) the end of the 2 hr densification heat treatment, b) electrochemical operation, and c) room temperature shutdown.

Effect of Parameters on Densification in Stack Structures

The sensitivity of the densification behavior to various parameters was also studied. The enhanced sintering using a smaller initial grain size was observed. The smaller grain sizes resulted in a faster sintering rate and improved final density for the center of the stack, but the density variation across the layer remained (Figure 8). The maximum relative density increased from 55% with the default 1.0 m grain to 66% with the 0.25 m initial grain size. The low densification of the corners/edges remained due to the outer frame's mechanical influence. The effect on contact layer stresses was small. The edge regions of high stress during the sintering stage were the same, but the high stress regions at operation and shutdown increased slightly. The greater volumetric shrinkage during densification resulted in higher tensile stresses at the rib ends after room temperature shutdown (Figure 9), while the large area exceeding the threshold was nearly the same. Reduced grain size of the precursor materials will likely be helpful to increase densification, but the stack stiffness characteristics may result in greater shutdown stresses. The present stack model did not consider further densification or material creep at the operating temperature representative of long term stack operations, but these potential transient deformations would also strongly affect the final shutdown stress state.

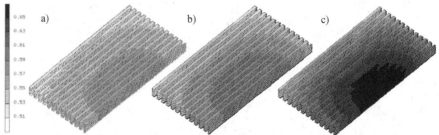

Figure 8. Contours of the final relative density for initial grain sizes of a) 0.75, b) 0.50, and c) 0.25 m.

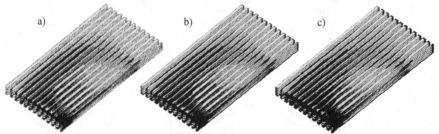

Figure 9. Contours of the maximum principal stress in excess of 15 MPa limit (shown in black) at room temperature shutdown for initial grain sizes of a) 0.75, b) 0.50, and c) 0.25 m.

The effect of mechanical preload was also investigated. Increased preload may induce greater compression on the contact layer during processing to increase its densification, and increased preload during thermal loading may help reduce tensile stress development. It was found that doubling the default 0.2 MPa preload to 0.4 MPa had only a small impact with only ~1.0% increased densification. This suggests that the small preload may be useful for ensuring components remain in physical contact during densification, but it does not provide any significant increase to the rate of densification. A preload of 2.0 MPa was required to see a more substantial impact by increasing the peak densification by ~9.8% and reducing the region of high stress upon shutdown (Figure 10c).

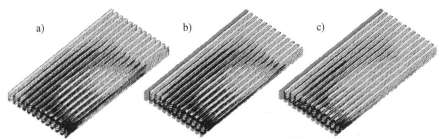

Figure 10. Contours of the maximum principal stress in excess of 15 MPa limit (shown in black) at room temperature shutdown for compressive preloads of a) 0.2, b) 0.4, and c) 2.0 MPa.

Finally, the effect of contact paste thermal expansion coefficient (TEC) was also evaluated. The thermal expansion coefficients of the material systems are well matched, but the influence of the TEC with +/- 5.0% variation was also tested. The influence on the sintered density was negligible due to the small thermal strain mismatch during the isothermal heat treatment, but reduced stresses were observed on shutdown with the reduced TEC (Figure 11a). This occurred because the reduced thermal shrinkage of the contact layer during shutdown relative to the rest of the stack imposes more compressive loading on the contact layer. This suggests that small changes to the TEC that are tolerable at the stack operating condition or use of a contact material that has a smaller TEC at low temperatures could be beneficial for increased thermal cycle tolerance.

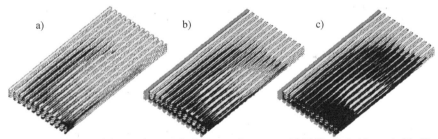

Figure 11. Contours of the maximum principal stress in excess of 15 MPa limit (shown in black) at room temperature shutdown for contact material TEC values of a) 95%, b) 100%, and c) 105% of the nominal value.

CONCLUSIONS

A constitutive model to simulate the sintering and densification of cathode contact materials was developed and implemented in the MSC Marc FEA code. The model was used to evaluate the densification behavior in specimens and stack geometries. The investigations provided the following observations and conclusions:

Good densification of candidate material systems under free sintering was predicted to be possible for temperatures less than 1000°C based on the experimental material data from dilatometry screening tests.

Small initial grain sizes less than 0.5 m were predicted to freely sinter to near full density.

Compressive preload increased the free sintering rate, but significant improvements were only realized once the applied stress was at or above the inherent sintering stress.

For the proposed two hour heat treatment at 900°C, the cathode contact layer densification in the generic stack geometry was poor overall, and density variations occurred across the cell due to the mechanical design and stiffness properties of the support frame.

Stresses in excess of the estimated failure strength occurred only at the ends of the cathode-side interconnect ribs during high temperature stack operation, but widespread stresses in excess of the estimated strength occurred during shutdown to room temperature.

Reduced initial grain size improved densification in the generic stack geometry with a small detrimental impact on the predicted stress state at operation and shutdown.

Increasing the stack compressive preload to 2.0 MPa increased the contact densification and reduced the contact layer tensile stresses occurring at shutdown.

Reduction of the TEC of the contact material could possibly be beneficial to reducing the high stresses in the contact layer at shutdown.

For future work, the constitutive parameters for the chosen contact material model should be evaluated more thoroughly, creep strains during long term operation should be included in the model, and the effect of component geometry on contact uniformity and predicted stresses should be characterized to improve the reliability of the contact layer in actual stack designs.

ACKNOWLEDGEMENTS
The authors would like to sincerely thank Gordon Xia and Jeff Stevenson for providing the experimental densification data and helpful discussions. The work summarized here was funded as part of the Solid-State Energy Conversion Alliance (SECA) Core Technology Program by the U.S. Department of Energy's National Energy Technology Laboratory (NETL).

REFERENCES
[1] J.W. Stevenson, G.G. Xia, Z. Lu, X. Li, Z. Nie, T. Oh, and J.D. Templeton, Development of Cathode Contact Materials for SOFC, presented at the 11th Annual SECA Workshop, Pittsburgh, PA, July 27-29, 2010. Accessed 12/28/2010 at http://www.netl.doe.gov/publications/proceedings/10/seca/Presentations/Stevenson1_Presentation.pdf.

[2] X. Montero, F. Tietz, D. Stover, M. Cassir, and I. Villareal, Comparative Study of Perovskites as Cathode Contact Materials between an $La_{0.8}Sr_{0.2}FeO_3$ Cathode and a Crofer22APU Interconnect in Solid Oxide Fuel Cells, *J. Power Sources*, **188**, 148-155 (2009).

[3] Z. Yang, G. Xia, P. Singh, and J. Stevenson, Electrical Contacts Between Cathodes and Metallic Interconnects in Solid Oxide Fuel Cells, *J. Power Sources*, **155**, 245-252 (2006).

[4] W.N. Liu, B.J. Koeppel, X. Sun, and M.A. Khaleel, Effect of Geometrical and Mechanical Properties of Various Components on Stresses of the Seals in SOFCs, in Advances in Solid Oxide Fuel Cells VI: *Ceram. Eng. Sci. Proc.*, **31**, 53-62 (2010).

[5] Olevsky EU, Theory of Sintering: From Discrete to Continuum, *Mater. Sci. Eng.*, **R23**, 41-100 (1998).

[6] Marc 2007 r1, Volume A: Theory and User Information, MSC Software Corporation, Santa Ana, CA (2007).

[7]A. Poirson, P. Decors, G. Caboche, and L.C. Dufour, A Dilatometric Study of the $La_{0.8}Sr_{0.2}MnO_3$ Sintering Behavior," *Solid State Ionics*, **99**, 287-295 (1997).

[8]S.-J.L. Kang, Sintering: Densification, Grain Growth and Microstructure, Elsevier (2005).

[9]M.A. Khaleel, B.J. Koeppel, W. Liu, K. Lai, K.P. Recknagle, E. Ryan, E.V. Stephens, and X. Sun, Modeling Tools for Solid Oxide Fuel Cell Analysis, presented at the 10th Annual SECA Workshop, Pittsburgh, PA, July 14-16, 2009. Accessed 12/28/2010 at http://www.netl.doe.gov/publications/proceedings/09/seca/presentations/Khaleel_Presentation.pdf.

OBSERVATIONS ON THE AIR ELECTRODE-ELECTROLYTE INTERFACE DEGRADATION
IN SOLID OXIDE ELECTROLYSIS CELLS

Michael Keane, Atul Verma, and Prabhakar Singh
Center for Clean Energy Engineering
Department of Chemical, Materials and Biomolecular Engineering
University of Connecticut
Storrs, CT, USA

ABSTRACT
        Delamination of the air electrode from the electrolyte in solid oxide electrolysis cells (SOEC) is a major contributor to long-term electrical performance degradation. In this study, the mechanism of this delamination has been investigated. Half cells consisting of 8 mol% YSZ electrolyte and LSM electrodes were fabricated by screen printing method and tested in air at atmospheric pressure and 840 °C for 100 hours under constant voltage ranging from 0 to 0.8 volts. Changes in microstructure and phase evolution at the electrode-electrolyte interfaces have been studied using scanning electron microscopy, energy dispersive x-ray spectroscopy, and x-ray diffraction. Hypotheses for the delamination mechanism have been developed that deal with the mechanical and chemical changes taking place at the air electrode-electrolyte interface.

INTRODUCTION
        Solid oxide electrolysis cells (SOEC) are electrochemical devices used for high temperature electrolysis (HTE) to produce hydrogen or syngas with no emissions of greenhouse gases. One envisioned HTE system uses the waste heat from a high temperature nuclear reactor to provide thermal and electrical energy to a SOEC for the production of hydrogen. At an operating temperature of 850 °C, achievable thermal to hydrogen efficiency is over 50%, which is much higher than that of a low temperature electrolysis system.[1]
        A solid oxide electrolysis cell (Figure 1) consists of a dense electrolyte and two porous electrodes. A mixture of water and hydrogen is delivered to the fuel electrode. The purpose of flowing hydrogen is to maintain a reducing atmosphere to stabilize the fuel electrode material. A voltage is applied between the electrodes to split the water into hydrogen gas and oxygen ions. The electrodes catalyze the electrochemical half reactions and provide conductive pathways for electrons to travel to the reaction sites. The oxygen ions diffuse through the electrolyte and exit as oxygen gas on the air electrode side. This mode of operation is essentially the reverse of that of a solid oxide fuel cell (SOFC).[2]
        In both SOECs and SOFCs, the electrolyte is typically yttria-stabilized zirconia (YSZ), and the fuel electrode is usually a nickel-YSZ composite. The air electrode is commonly lanthanum strontium manganite (LSM) or a LSM-YSZ composite[2]. SOECs are stacked in series to provide a higher rate of hydrogen production. The cells in a stack are separated by conductive interconnects that are usually made of stainless steel.[3]
        The voltage applied to each SOEC in a stack is typically around 1.3 volts, or thermal neutral voltage. At this voltage, the thermal energy consumed by the electrochemical reactions is equal to the thermal energy produced due to electrical resistance (ohmic heating) in the cell. Higher voltages are avoided because the additional ohmic heating could cause overheating of the SOEC stack. The open circuit (zero current) voltage (OCV) is around 0.9 to 1 volt for each cell, depending on temperature and gas composition. Thus, SOECs are typically run at 0.3 to 0.4 volts above OCV.[4]

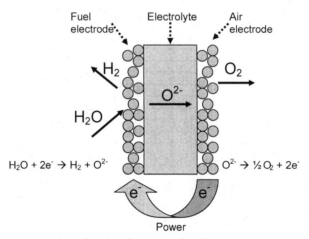

Figure 1 Schematic and half-reactions of a solid oxide electrolysis cell.

Figure 2 shows a sample voltage – current density curve for a solid oxide cell. The left half the graph (negative current) represents electrolysis mode, where the cell voltage is larger than the OCV. The right half of the graph (positive current) represents fuel cell mode, where the cell voltage is smaller than OCV. The flat line at 1.3 volts represents thermal neutral voltage.

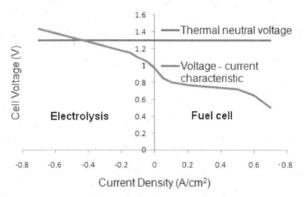

Figure 2 Sample voltage – current characteristic curve of a solid oxide cell in both electrolysis and fuel cell mode. Thermal neutral voltage is approximately 1.3 volts.

One of the most important developmental barriers in SOEC technology is that of long-term degradation. Cell degradation is typically responsible for a 20-30% reduction in hydrogen production rate in 1000 hour tests with SOEC stacks[4]. Clearly, such degradation is too severe for SOECs to become commercially viable with the current state of technology.

There are many issues thought to be the cause of this degradation. Chromium species may evaporate from the steel interconnects and deposit on the electrodes, reducing the triple phase boundary (TPB) length.[5] Nickel from the fuel electrode can evaporate or coarsen, also reducing TPB length.[6] The bipolar conditions (oxidizing and reducing atmospheres on opposite sides) experienced by the interconnects can cause corrosion and scaling.[7] Based on the observations from the limited long-term SOEC testing, the largest contributor to cell performance degradation is believed to be delamination of the air electrode from the electrolyte.[3,8-11] These issues are summarized in Figure 3.

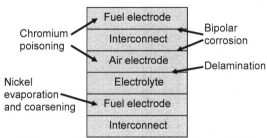

Figure 3 Summary of degradation-related issues facing SOEC stacks.

Although exact reason for the origin of such delamination is not clear, a model for mechanical failure has been proposed related to the mismatch between the greater ability of zirconia to release oxygen and lesser ability of the electrode material, such as LSM, to conduct oxide ions away. Under these conditions, defect at the solid–solid interface between the two materials serve as a nucleation point where oxygen can be released after the electron is transferred to the electrode. The high oxygen pressure may cause enough interfacial tension to detach the air electrode from the electrolyte. The delamination in this case is mechanical in nature.[12]

It should be noted that in SOFCs, oxygen is entering the electrolyte from the air electrode side. Therefore, SOFCs will not experience high oxygen pressures at the air electrode-electrolyte interface. Consequently, air electrode delamination in solid oxide cells has only been reported in electrolysis operation, in which oxygen evolves at the air electrode-electrolyte interface.

Air electrode delamination has been investigated in this study using electrochemical tests with a half cell configuration. Delamination mechanism hypotheses have been developed from observations of chemical and morphological changes at the air electrode-electrolyte interfaces of tested cells.

EXPERIMENTAL

The degradation at the oxygen electrode-electrolyte interface was studied using half cells, with the air electrode material on both sides of the electrolyte. This allows for simplification of the experiments because air is flowed on both sides of the cell, without any high temperature gas sealing requirements. As voltage is applied, oxygen ions are pumped through the cell, and the cell electrochemical behavior is monitored over several days.

Cells were fabricated from $(ZrO_2)_{0.92}(Y_2O_3)_{0.08}$ (YSZ) electrolyte substrates obtained from ENrG Inc, Buffalo, NY. $La_{0.8}Sr_{0.2}MnO_{3-x}$ (LSM) electrode paste from Nextech Materials was subsequently screen printed on both sides of the YSZ disc using a semi-automated screen printer with 105 mesh screen, then sintered in air for 2 hours at 1200 °C. The air electrodes (oxygen evolving side) were made smaller than the cathodes (oxygen entering side) in order to assure that oxygen would evolve over the entire surface area of the air electrode. Silver mesh current collectors were attached to each electrode using silver paste. Completed half cells are shown in Figure 4.

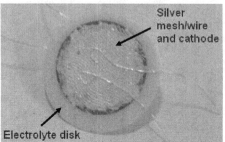

Figure 4 Untested half cell (cathode side) consisting of a YSZ electrolyte disk, screen-printed LSM electrodes, and silver mesh and wires attached with silver paste.

Each cell was placed on top of an alumina tube inside a furnace and the silver wires ran down the sides of the tube to beneath the furnace (Figure 5). Leads from a Bio-Logic VMP2 multi-channel potentiostat were attached for voltage application and impedance measurements during testing.

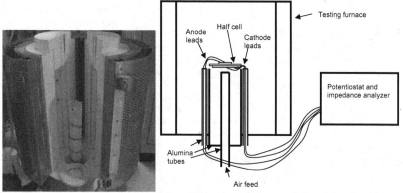

Figure 5 Photo of half-cell on top of alumina tube inside open furnace (left); schematic of the same setup (right).

The cells were heated to the operating temperature of 840 °C in flowing air. Using the potentiostat, a constant voltage was then applied for 100 hours and the cell current was monitored throughout the test. As previously mentioned, typical SOECs are run at 0.4 volts above OCV for thousands of hours. In the current study, cells were tested using high voltages (up to 0.8 volts) for the observation of accelerated degradation in a shorter time than in actual SOEC stacks. The electrode and electrolyte resistances were measured as a function of time by superimposing impedance spectroscopic measurements with 10 mV ac excitation voltage. Impedance measurements were taken at an interval of every four hours.

Electrolyte and electrode interfaces developed at both the air electrode and cathode were subsequently examined using scanning electron microscopy (SEM). Electrode–electrolyte interfaces were also examined after the dissolution of the LSM electrode to understand the electrolyte surface morphology. X-ray diffraction (XRD) analysis was also conducted to understand interfacial compound formation.

RESULTS

After cool down, the air electrode (oxygen evolution electrode) of the 0.8 volt bias cell was found completely delaminated and detached from the electrolyte (Figure 6). This delamination did not occur in cells tested with lower voltages, nor did it occur on the cathode sides of any cells.

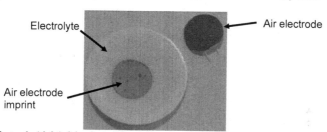

Figure 6 Air electrode (right) delaminated from electrolyte (left) after 100 hours of applying 0.8 volts. The dark area in the center of the electrolyte is the imprint of the air electrode after delamination.

Table I summarizes the delamination results from five of the tests at various voltages. Also shown are the relative increases in electrolyte and electrode resistances after 70 hours of testing. A considerable increase in electrolyte resistance is only seen when 0.8 volts is applied. The rate of electrode degradation does not correlate as well with voltage, which is likely due to cell-to-cell variation of electrode characteristics.

Table I Testing conditions and relative resistances after 70 hours from various tests with LSM electrodes.

Cell ID#	DC Bias (V)	Electrolyte resistance	Electrode resistance	Degree of delamination
5	0	1.1	2.4	None
28	0.3	1.05	0.77	None
27	0.5	1.1	0.9	None
4	0.8	77.7	14.1	Complete

The post-test microscopic examination revealed different microstructural features present on the two electrode surfaces. The electrode-electrolyte contact interfaces (active areas) were exposed by dissolving the LSM phase in hydrochloric acid.

Figure 7 shows the electrolyte surface features developed under various conditions. Figure 7a shows the initial electrolyte surface. The surface shows a typical dense equiaxed structure with localized isolated pores.

Figure 7b shows the morphology developed after screen printing and sintering of the electrode. As is evident, the electrolyte maintains its dense granular structure with evidence of a peripheral impression of the air electrode in the form of elevated ridges. These ridges are only present in the active area (interface with LSM) on the left side of the image. They are due to the nucleation of epitaxial (Mn)-YSZ at the LSM-YSZ triple phase boundary (TPB). This nucleation is followed by surface diffusion of Zr^{4+} and Y^{3+} cations towards the TPB, as explained in detail by Mitterdorfer et al.[13]

Figure 7c represents the air electrode side YSZ surface after applying 0.8 volts for 100 hours. The YSZ surface in this case clearly shows several drastic morphological changes. Pores have formed at the grain boundaries in the active area (interface with LSM) on the left side of the image. These pores are thought to be a direct result of high pressure oxygen evolving from the electrolyte. Circular pits have formed on the YSZ surface, which may also be due to high pressure effects. There appears to have been a reaction at the interface which left behind a network of small particles that covers most of the active YSZ surface. These changes are present throughout the entire air electrode-electrolyte interfacial area.

For comparison, the cathode surface of 0.8 volt cell is also shown (Figure 7d). The only morphological changes observed are the formation of elevated ridges from the LSM as seen in the untested cell interfaces (Figure 7b).

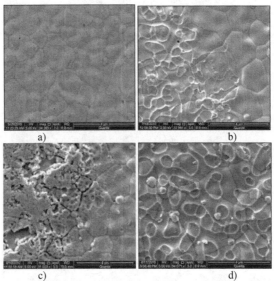

Figure 7 Comparison of electrolyte surface morphologies developed after dissolving LSM in hydrochloric acid. Each image shows approximately 10 μm × 8 μm; a) Exposed surface of as received electrolyte disc; b) Surface in contact with air electrode with no electrochemical testing; c) Surface in contact with air electrode after applying 0.8 volts for 100 hours; d) Surface in contact with cathode after applying 0.8 volts for 100 hours.

Figure 8 shows x-ray diffraction patterns of the LSM air electrode-side electrolyte surfaces after etching in hydrochloric acid. Patterns were taken for an untested cell, a cell tested at open circuit voltage (OCV) for 100 hours, and a cell tested at 0.8 volts for 100 hours. The dots identify the positions of lanthanum zirconate ($La_2Zr_2O_7$) diffraction peaks, which were observed only in the 0.8 volt cell.

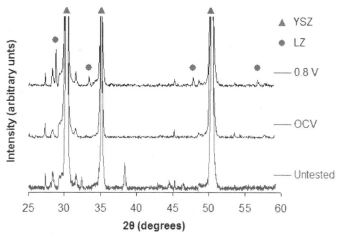

Figure 8 X-ray diffraction patterns of LSM air electrode-side electrolyte surfaces after etching in hydrochloric acid. Top: untested; middle: tested at OCV; bottom: tested at 0.8 volts; triangles: YSZ diffraction peaks[14]; dots: La$_2$Zr$_2$O$_7$ diffraction peaks[15].

DISCUSSION

Microscopy and x-ray diffraction indicate that major electrode-electrolyte interfacial changes are limited to the air electrode side of cells tested under high voltage. These morphological and chemical changes do not take place on the cathode side, nor do they take place in cells not tested under voltage. This suggests that electrolyte morphological changes, lanthanum zirconate formation, and eventual air electrode delamination are caused by the evolution of oxygen at the air electrode-electrolyte interface.

To explain the mechanism of delamination, two types of hypotheses have been formed. The first involves "mechanical failure" at the interface due to high pressure, and the second type involves "chemical failure" due to phase formation or stoichiometric changes in LSM near the interface.

Mechanical Failure

The maximum interfacial oxygen partial pressure attainable depends on the electrical and ionic conductivities of the electrolyte and electrode, as well as their thicknesses, the temperature, and the applied voltage. Calculated maximum oxygen pressure for three common electrode materials are given in Table II. Also listed are the electronic and ionic conductivities of the materials. The parameters used are the cell testing conditions with 0.8 volts applied.

Table II. Electrode conductivities and calculated maximum oxygen pressure at air electrode-YSZ electrolyte interfaces.

Electrode material	Electronic conductivity (S/cm)	Ionic conductivity (S/cm)	Max pressure (atm)
LSM	126	$7.7 \cdot 10^{-8}$	$7.3 \cdot 10^{22}$
LSM-YSZ	42	$8.0 \cdot 10^{-4}$	$9.7 \cdot 10^{19}$
LSCF	100	0.12	2.4

Since LSM has low ionic conductivity[16] oxygen gas could concentrate in closed pores at the air electrode-electrolyte interface and attain very high pressures. These high pressures could be responsible for the delamination observed in the current study. On the other hand, due to the high ionic conductivity of LSCF, oxygen pressure buildup will be minimal, so LSCF or other mixed conducting air electrodes should not delaminate due to interfacial oxygen pressure.

Chemical Failure

Hypotheses have been formed that are related to the chemical changes leading to phase change or shrinkage behavior in the LSM air electrodes when exposed to high oxygen partial pressure. The first proposed model deals with phase changes driven by the high oxygen pressures. Lanthanum zirconate formation is accelerated by high oxygen pressures by reaction (1).[17]

$$LaMnO_3 + ZrO_2 + 0.25\ O_2 \rightarrow 0.5\ La_2Zr_2O_7 + MnO_2 \tag{1}$$

The resulting excess manganese will then diffuse into the YSZ.[13] The various phases formed will presumably have different coefficients of thermal expansion, and therefore local temperature gradients could causes stress buildup at the air electrode-electrolyte interface, leading to delamination.

Another hypothesis refers to the chemical changes in the LSM lattice in high oxygen partial pressure due to the well known hyperstoichiometric behavior of $La_xSr_{1-x}MnO_{3+\delta}$ with an oxygen excess $(\delta > 0)$[18]. This behavior results in metal ion vacancies and shrinkage of the LSM lattice[19]. It is hypothesized that lattice shrinkage near the air electrode-electrolyte interface could cause local stresses and eventual air electrode detachment.

CONCLUSIONS

This study has demonstrated a positive correlation between high applied voltage, air electrode delamination, morphology changes at the air electrode-electrolyte interface, and cell degradation rate. The mechanism of LSM delamination is hypothesized to be based on the oxygen pressure buildup at the air electrode-electrolyte interface. The mechanism may be mechanical or chemical in nature. The electrode may experience mechanical separation as a direct result of high interfacial oxygen pressures. These high pressures could also cause phase changes at the LSM-YSZ interface or stoichiometric changes in the LSM leading to shrinkage behavior. Both of these chemical changes could cause local stresses at the air electrode-electrolyte interface and eventual delamination.

ACKNOWLEDGEMENTS

The authors would like to thank Idaho National Laboratory for financial support of this work. Also acknowledged are Drs. Manohar Sohal, James O'Brien, and Carl Stoots from INL for helpful discussion, as well as Drs. Sanjit Bhowmick and Manoj Mahapatra from the Center for Clean Energy Engineering for technical discussion and assistance in surface analysis.

REFERENCES

[1]J.E. O'Brien, M.G. McKellar, E.A. Harvego, and C.M. Stoots. High-temperature electrolysis for large-scale hydrogen and syngas production from nuclear energy – summary of system simulation and economics analyses. *International Journal of Hydrogen Energy* **35**, 4808–4819 (2010).
[2]C. Yang, A. Coffin, and F. Chen. High temperature solid oxide electrolysis cell employing porous structured $(La_{0.75}Sr_{0.25})_{0.95}MnO_3$ with enhanced oxygen electrode performance. *International Journal of Hydrogen Energy* **35**, 3221–3226 (2010).

[3] J.R. Mawdsley, J.D. Carter, A.J. Kropf, B. Yildiz, and V.A. Maroni. Post-test evaluation of oxygen electrodes from solid oxide electrolysis stacks. *International Journal of Hydrogen Energy* **34**, 4198–4207 (2009).

[4] J.E. O'Brien, C.M Stoots, J.S. Herring, and J.J. Hartvigsen. Performance of planar high-temperature electrolysis stacks for hydrogen production from nuclear energy. *Nuclear Technology* **158**, 118–131 (2007).

[5] X. Chen, Y. Zhen, J. Li and S.P. Jiang. Chromium deposition and poisoning in dry and humidified air at $(La_{0.8}Sr_{0.2})_{0.9}MnO_{3-x}$ cathodes of solid oxide fuel cells. *International Journal of Hydrogen Energy* **35**, 2477–2485 (2010).

[6] L. Holzer, B. Iwanschitz, Th. Hocker, B. Münch, M. Prestat, D. Wiedenmann, U. Vogt, P. Holtappels, J. Sfeir, A. Mai, and Th. Graule. Microstructure degradation of cermet anodes for solid oxide fuel cells: Quantification of nickel grain growth in dry and in humid atmospheres. *Journal of Power Sources* **196**, 1279–1294 (2011).

[7] Z. Yang, G. Xia, M.S.Walker, C. Wang, J.W. Stevenson, P. Singh. High temperature oxidation/corrosion behavior of metals and alloys under a hydrogen gradient. *International Journal of Hydrogen Energy* **32**, 3770–3777 (2007).

[8] A. Kaiser, E. Monreal, A. Koch, and D. Stolten. Reactions at the Interface $La_{0.5}Ca_{0.5}MnO_3$-YSZ/Al_2O_3 under Anodic Current. *Ionics* **2**, 184–189 (1996).

[9] A. Momma, T. Kato, Y. Kaga, and S. Nagata. Polarization Behavior of High Temperature Solid Oxide Electrolysis Cells (SOEC). *Journal of the Ceramic Society of Japan* **105** [5], 369–373 (1997).

[10] J. Guan, N. Minh, B. Ramamurthi, J. Ruud, J. Hong, P. Riley, and D. Weng. High Performance Flexible Reversible Solid Oxide Fuel Cell. GE Global Research Center, 2004-2006.

[11] H. Lim and A.V. Virkar. A study of solid oxide fuel cell stack failure by inducing abnormal behavior in a single cell test. *Journal of Power Sources* **185**, 790–800 (2008).

[12] A.V. Virkar, J. Nachlas, A.V. Joshi, and J. Diamond. Internal Precipitation of Molecular Oxygen and Electromechanical Failure of Zirconia Solid Electrolytes. *Journal of the American Ceramic Society* **73** [11], 3382–3390 (1990).

[13] A. Mitterdorfer and L.J. Gauckler. $La_2Zr_2O_7$ formation and oxygen reduction kinetics of the $La_{0.85}Sr_{0.15}Mn_yO_3$, $O_2(g)|YSZ$ system. *Solid State Ionics* **111**, 185–218 (1998).

[14] G. Laukaitis, J. Dudonis, and D. Milcius. YSZ thin films deposited by e-beam technique. *Thin Solid Films* **515**, 678–682 (2006).

[15] J.Y. Li, H. Dai, Q. Li, X.H. Zhong, X.F. Ma, J. Meng, and X.Q. Cao. Lanthanum zirconate nanofibers with high sintering-resistance. *Materials Science and Engineering B* **133** 209–212 (2006).

[16] W. Huang, S. Gopalan, and U. Pal. Measurement of partial oxygen ion conductivity of Sr-doped lanthanum manganite. *Journal of Power Sources* **173**, 887–890 (2007).

[17] M. Chen. Understanding the Thermodynamics at the LaMnO3-YSZ Interface in SOFC. Ph.D. Dissertation. ETH Zurich, 2005.

[18] K. Nakamura. The defect chemistry of $La_{1-\Delta}MnO_{3+\delta}$. *Journal of Solid State Chemistry* **173**, 299–308 (2003).

[19] S. Miyoshi, A. Kaimai, H. Matsumoto, K. Yashiro, Y. Nigara, T. Kawada, and J. Mizusaki. In situ XRD study on oxygen-excess LaMnO3. *Solid State Ionics* **175**, 383–386 (2004).

Fuel Reforming

CARBON DIOXIDE REFORMING OF METHANE FOR SOLID OXIDE FUEL CELLS

Mitsunobu Kawano, Hiroyuki Yoshida, Koji Hashino, and Toru Inagaki
The Kansai Electric Power Company, Inc.
11-20 Nakoji 3-chome, Amagasaki, Hyogo 661-0974, Japan

ABSTRACT
    Carbon dioxide reforming of methane on nickel (Ni)-ceria-based oxide cermet anode of practical size solid oxide fuel cells (SOFCs) was carried out under various conditions. The cell performance and the gaseous composition of the fuel were measured under various conditions. Stable cell performance was observed with CH_4 / CO_2 ratios being 1/4, 1/3, 1/2 at 750℃. The results of gas analysis reveal that methane was effectively converted to hydrogen and carbon monoxide, and that hydrogen and carbon monoxide were effectively used as fuel species for eletrochemical oxidation reactions. The results show that the anode works effectively for the carbon dioxide reforming, resulting in the stable SOFC performance. It is concluded that carbon dioxide reforming attained sufficient level of conversion for power generation with CH_4 / CO_2 ratios being 1/4, 1/3, 1/2 at 750℃ on Ni-ceria-based oxide cermet anode of practical size SOFCs.

INTRODUCTION

    Solid oxide fuel cells (SOFCs) are expected to be a new attractive technology for electrical power generation because of their high electrical efficiency. In general, they must operate at high temperatures near 1000°C. But such high operating temperatures cause many serious problems such as physical and chemical degradations of the SOFCs component materials. Then, intermediate temperature (600-800℃) SOFCs (IT-SOFCs) have been drawing a great deal of attention. The Kansai Electric Power Company Inc. (KEPCO) has been developing IT-SOFCs with nickel (Ni)-ceria-based oxide cermet anode. Ni-ceria-based oxide cermet anode gives high performance due to synergistic effect of enlarging reaction area and increasing paths for ionic and electronic conduction[1].
    Another attractive feature of SOFCs is their capacity of internal reforming operation using hydrocarbon fuel because of their higher operating temperatures than the other types of fuel cells[2, 3]. Reforming operation method is generally classified in two ways, which are steam reforming (Reaction 1) and carbon dioxide reforming (Reaction 2). Steam reforming operation has been generally used for a practical SOFC application[4-7]. Carbon dioxide reforming operation, however, has not been adopted as a practical SOFC application. Because steam generator is not necessary and higher electrical efficiency is expected for carbon dioxide reforming operation, it should have more advantage than steam reforming operation. Moreover, methane and carbon dioxide are the main constituents of biogas produced by anaerobic biological waste treatment[8, 9]. Then, investigation of the carbon dioxide reforming operation of SOFCs with methane fuel is considered to be very important for the development of biogas fueled SOFCs.
    In this study, carbon dioxide reforming of methane on nickel (Ni)-ceria-based oxide cermet anode of practical size SOFCs under various conditions was carried out to reveal the electrochemical and the reforming activities for its operation of SOFCs. In addition, external and direct internal reforming operations were compared to clarify its operation in detail. Moreover, carbon dioxide reforming operation was compared with steam reforming operation we have reported previously[10].

$$CH_4 + H_2O \rightarrow CO + 3H_2 \qquad \Delta H^0_{298} = 206 \text{ kJ mol}^{-1} \qquad (1)$$
$$CH_4 + CO_2 \rightarrow 2CO + 2H_2 \qquad \Delta H^0_{298} = 247 \text{ kJ mol}^{-1} \qquad (2)$$

EXPERIMENTAL

Cell fabrication process

Cell fabrication was carried out as the similar method we have reported previously[11, 12]. Optimized NiO-$(Ce,Sm)O_2$. (SDC) composite particles synthesized by spray pyrolysis were used as raw materials of the anode[11, 12]. NiO-SDC composite particles were reduced into Ni-SDC cermet after the processes of sintering and reducing under a hydrogen atmosphere. $La_{0.8}Sr_{0.2}Ga_{0.8}Mg_{0.15}Co_{0.05}O_3$. (LSGMC) with 200 μm thickness and $Sm_{0.5}Sr_{0.5}CoO_{3-\delta}$ (SSC) were selected as an electrolyte and a cathode, respectively. The diameter of electrodes was 120 mm.

Electrochemical characterizations

As schematically shown in Figure 1, air and fuel were supplied to the center of the cathode and the anode, respectively. This single cell stack adopted the seal-less structure in which the unspent residual fuel and air were exhausted at the outer surface of the single cell stack. The electrochemical measurement system is schematically drawn in Figure 2. The cell performance was tested at 750, 700, and 650℃. Air was supplied to the cathode at a flow rate of 15 ml cm $^{-2}$min^{-1}. Under the condition of changing operation temperature, hydrogen and nitrogen were supplied to the anode to keep a reducing atmosphere. In power generation, methane and carbon dioxide with various ratios were also supplied to the anode for the carbon dioxide reforming. The flow rate of methane was fixed at 0.75 ml cm^{-2}min^{-1} and the flow rates of carbon dioxide were changed at 3.00, 2.25, and 1.50 ml cm^{-2}min^{-1} to realize CH_4 / CO_2 ratios being 1/4, 1/3, and 1/2, respectively. In comparison with carbon dioxide reforming, steam reforming was also carried out. For the steam reforming, methane and steam were supplied to the anode with flow rates of 0.75 and 2.25 ml cm^{-2}min^{-1}, respectively, to realize CH_4 / H_2O ratio being 1/3. For comparison, dry hydrogen was also supplied to the anode at a flow rate of 3.00 ml cm^{-2}min^{-1}. The amount of supplied hydrogen was determined to be four times as much as that of supplied methane because 1 mol of methane is converted to 2 mol of hydrogen and 2 mol of carbon monoxide when carbon dioxide reforming reaction of methane proceeded (Reaction 2). For the external carbon dioxide reforming, methane and carbon dioxide were supplied to the pre-reformer containing $Ru-Al_2O_3$ reforming catalyst (27 cm^3), prior to the introduction to the anode. The temperature of the pre-reformer was set at the same of the operation temperature (750, 700, and 650℃). In the case of the direct internal carbon dioxide reforming, methane and carbon dioxide were directly supplied to the anode through the by-pass line of the pre-reformer. The composition of the gaseous species in the reformate except for water vapor was measured by micro-gas chromatograph (Varian, SP-4900). As shown in Figure 3, the anode separator had five points for sampling the gaseous species in the radial direction, which were 0 mm, 14 mm, 28 mm, 42 mm, and 56 mm, respectively. These values mean the distance from the center hole of the separator. For the electrochemical characterization, I-V characteristics were measured with increasing current density. AC impedance was measured in the frequency range from 100 kHz to 0.1 Hz with an AC amplitude of 17.7 mA cm^{-2}. In this paper, we denote external carbon dioxide reforming as ECDR, also denote direct internal carbon dioxide reforming as DICDR.

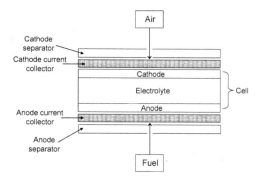

Figure 1. Schematic view of the cross-section of the SOFC single cell stack.

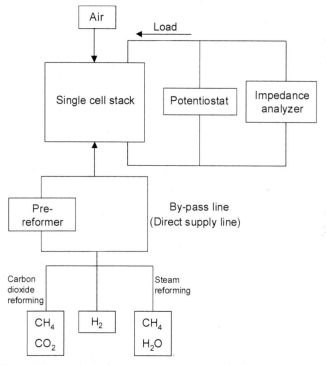

Figure 2. Electrochemical measurement system for SOFC power generation.

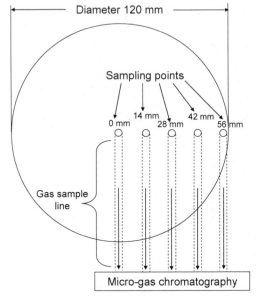

Figure 3. Schematic view of the anode separator for gas analysis.

RESULTS AND DISCUSSION

Temperature dependence of SOFC power generation

SOFC power generation with ECDR and DICDR of methane was carried out at various temperatures of 750, 700, and 650℃. Figure 4 shows *I-V* characteristics of the cell at various temperatures with ECDR, DICDR and dry hydrogen fueled operations. As shown in Figure 4 (a), it is indicated that *I-V* characteristics of the cells were almost the same irrespective of the operation mode of ECDR and DICDR of methane at 750℃. Figure 4 (a) also shows that *I-V* performances of the cells with current density around 0.4 A cm^{-2} were almost the same irrespective of dry hydrogen and carbon dioxide-reformed methane fuels. These results indicate that methane fuel was sufficiently reformed on Ni-SDC anode regardless of the operation runs for ECDR and DICDR at 750℃. In the operation at 700℃, the *I-V* performance with high current density around 0.4 A cm^{-2} with DICDR operation was slightly lower than that with ECDR operation and dry hydrogen fueled operation (Figure 4 (b)). It is suggested that the *I-V* curves depended on supplied fuel conditions. *I-V* performance was significantly deviated among the supplied fuels at 650℃ (Figure 4 (c)). From Figure 4 (c), it is obvious that the *I-V* performances were the sequence of DICDR < ECDR < H$_2$ fuel; i.e., the H$_2$ fuel led to the highest current density at a given voltage and direct supply of methane led to the lowest one. These results indicate that methane conversions for the carbon dioxide reforming on Ni-SDC anode at 700 and 650℃ were incomplete than that at 750℃. The incomplete conversion of reforming resulted in the relatively large unevenness of each *I-V* performance.

AC impedance spectra were also measured for the cell. As described in our previous paper[15], the impedance of the overall cell reflected mainly the differences of the anode performance because the polarization loss due to the cathode was small[14]. The impedance spectra for the cell at various temperatures of 750, 700, and 650℃ under the open circuit voltage condition are shown in Figure 5. These impedance spectra were composed of high- and low-frequency semicircles which can be attributed to the resistances due to the activation and concentration polarization, respectively. From Figure 5, the high-frequency semicircles were almost same for ECDR and DICDR operations at each operation temperature. These results indicate that the resistances due to the activation polarization with ECDR and DICDR operations were almost same at each operation temperature. Meanwhile, the low-frequency semicircles with DICDR operation became gradually larger than those with ECDR operation as operation temperature decreased. From these results, it is considered that the resistances due to the concentration polarization with DICDR operation became larger than those with ECDR operation as the operation temperature decreased because of the insufficient conversion of methane at Ni-SDC anode with lower operation temperatures. These results from the impedance analysis show good agreement with those from the *I-V* characteristics. As a result, DICDR operation is strongly influenced by the operation temperature. DICDR attained sufficient level of conversion for power generation with methane at 750℃, while DICDR gave insufficient conversion at 700 and 650℃.

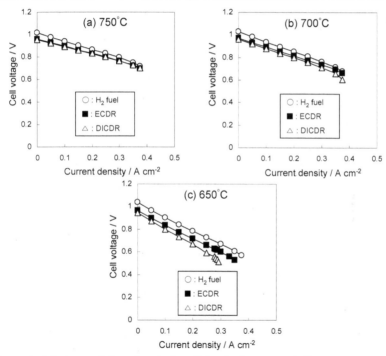

Figure 4. *I-V* characteristics of a single cell by feeding external direct internal carbon dioxide-reformed methane fuels (CH$_4$/CO$_2$ =1/4) and dry hydrogen fuel at (a) 750°C, (b) 700°C, and (c) 650°C.

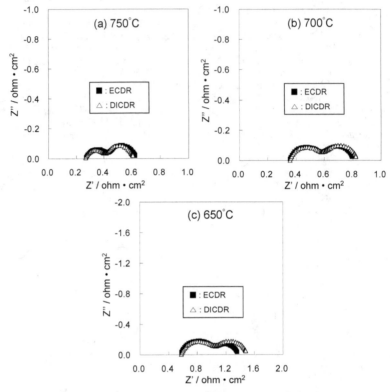

Figure 5. Impedance spectra of a single cell with the open circuit voltage condition by feeding external and direct internal carbon dioxide-reformed methane fuels (CH₄/CO₂ =1/4) at (a) 750, (b) 700, and (c) 650°C.

Dependence of fuel condition with CH_4 / CO_2 ratio for SOFC power generation

As shown in Figure 6, SOFC power generation with ECDR and DICDR of methane was also carried out at 750€ using three kinds of fuel conditions; CH_4/CO_2 = 1/4, 1/3, and 1/2. I-V characteristics of the cells were almost the same irrespective of the operation mode of ECDR and DICDR for CH_4/CO_2 = 1/4, 1/3, and 1/2 at 750€. These results indicate that methane fuel was sufficiently reformed on Ni-SDC anode regardless of the operation runs for ECDR and DICDR at 750€. DICDR reaction is suggested to proceed sufficiently from Figure 6, then, the comparison of I-V characteristics with dry hydrogen fuel was carried out (Figure 7). The amount of supplied hydrogen was determined to be four times as much as that of supplied methane because 1 mol of methane is converted to 2 mol of hydrogen and 2 mol of carbon monoxide when carbon dioxide reforming reaction of methane proceeded (Reaction 2). It is assumed that only hydrogen and carbon monoxide are used as fuel species for eletrochemical oxidation reactions. From Figure 7, open circuit voltage gradually increased as CH_4 / CO_2 ratios increased. As CH_4 / CO_2 ratio increased, carbon dioxide, which was the oxygen source in the supplied fuel, was decreased. Then, it is reasonable that open circuit voltage increased as CH_4 / CO_2 ratio increased because partial pressure of oxygen at the anode decreased. Under the conditions of high current densities around 0.3-0.4 A cm^{-2}, however, cell voltages were almost the same irrespective of the supplied fuel at 750℃. These results suggest that the polarization loss increased slightly as CH_4 / CO_2 ratio increased under the high current densities. AC impedance spectra were also measured for the cell. The impedance spectra for the cell at various methane to carbon dioxide ratios (CH_4/CO_2 = 1/4, 1/3, 1/2) are shown in Figure 8. This figure with the discharged conditions of 0.3 A cm^{-2} at 750℃ reveals that both high- and low-frequency semicircles became slightly smaller with decreasing CH_4 / CO_2 ratio. It is suggested that methane is effectively reformed by carbon dioxide at lower CH_4 / CO_2 ratio. These results from the impedance analysis showed good agreement with those from the data of I-V characteristics. Judging from Figures 7 and 8, it is considered that methane was effectively reformed as carbon dioxide increased. However, it can be considered that methane fuel was almost reformed on Ni-SDC anode for the DICDR with CH_4 / CO_2 ratios being 1/4, 1/3, 1/2 at 750℃. These phenomena of almost complete reforming were also observed in the steam reforming operation[10].

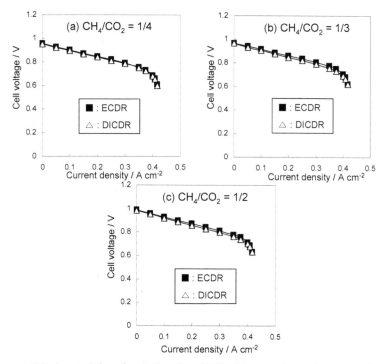

Figure 6. I-V characteristics of a single cell at 750℃ with external and direct internal carbon dioxide reforming of methane; CH_4/CO_2 = (a)1/4, (b)1/3, and (c)1/2.

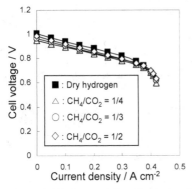

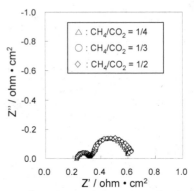

Figure 7. *I-V* characteristics of a single cell by feeding various direct internal carbon dioxide-reformed methane fuels and dry hydrogen fuel at 750℃.

Figure 8. Impedance spectra of a single cell with the discharged conditions of 0.3 A cm^{-2} at 750℃ using direct internal carbon dioxide-reformed methane fuel.

Comparison between carbon dioxide reforming and steam reforming for SOFC power generation

The comparison of the cell performance between direct internal steam-reformed methane (CH$_4$ /H$_2$O = 1/3) and carbon-dioxide-reformed methane (CH$_4$ /CO$_2$ = 1/3) at 750℃ was also carried out. The data of *I-V* characteristics and impedance spectra with direct internal steam reforming were previously reported[10]. For the carbon dioxide reforming, 1 mol of methane is converted to 2 mol of hydrogen and 2 mol of carbon monoxide when reforming reaction of methane proceeded (Reaction 2). On the other hand, 1 mol of methane is converted to 3 mol of hydrogen and 1 mol of carbon monoxide for steam reforming (Reaction 1). Total fuel species (hydrogen and carbon monoxide) were same for both carbon dioxide and steam reforming, but the ratios of hydrogen and carbon monoxide were different. Their thermodynamically calculated partial pressures of oxygen at the anode and their theoretical terminal voltages were almost the same. Although their theoretical terminal voltages were almost the same, open circuit voltage for the steam reforming was slightly higher than that for the carbon dioxide reforming, as shown in Figure 9. The reason was not obvious. However, the voltage for the carbon dioxide reforming at high current densities around 0.3-0.4 A cm^{-2} was higher than that for the steam reforming. Impedance spectra with the discharged conditions of 0.3 A cm^{-2} shown in Figure 10 also reveals that both high- and low-frequency semicircles for the carbon dioxide reforming were smaller than those for the steam reforming. These results indicate that the carbon dioxide reforming operation provided for the smaller resistance at high current densities. Although it was expected that electrochemical activity of hydrogen was higher than that of carbon monoxide, this assumption conflicted with these results. It is considered that steam might deteriorate Ni-SDC anode activity during the steam reforming operation. As a result, the carbon dioxide operation is considered to be useful compared with the steam reforming operation for practical size Ni-SDC anode.

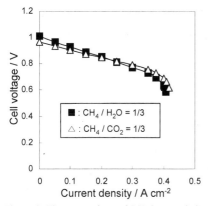

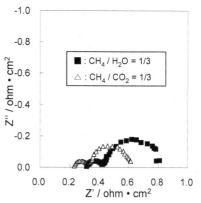

Figure 9. The comparison of *I-V* characteristics for a single cell between direct internal steam-reformed methane (CH$_4$ /H$_2$O = 1/3) and carbon dioxide-reformed methane (CH$_4$ /CO$_2$ = 1/3) at 750℃.

Figure 10. The comparison of impedance spectra for a single cell between direct internal steam-reformed methane (CH$_4$ /H$_2$O = 1/3) and carbon dioxide-reformed methane (CH$_4$ /CO$_2$ = 1/3) with the discharged conditions of 0.3 A cm^{-2} at 750℃.

Gas analysis in the radial direction of a single cell

As shown in Figure 3, the gaseous composition was measured at five sampling points in the radial direction which were 0 mm, 14 mm, 28 mm, 42 mm, and 56 mm, respectively. These values mean the distance from the center of the cell. Figure 11 shows the results of gas analysis for methane, hydrogen, and carbon monoxide in the radial direction by feeding carbon dioxide-reformed methane (CH$_4$ /CO$_2$ = 1/3) with the conditions of open circuit voltage and 0.3 A cm^{-2} at 750℃ after 2000 h durability test. Water vapor could not be analyzed because of the restriction of micro-gas chromatograph equipment. As shown in Figure 11 with the open circuit voltage condition, methane and carbon dioxide were gradually converted to hydrogen and carbon monoxide in the radial direction. These results revealed that the carbon dioxide reforming reaction (Reaction 2) proceeded effectively on Ni-SDC anode under the conditions of CH$_4$ /CO$_2$ ratio being 1/3 at 750℃. From Figure 11, the comparison between open circuit voltage and 0.3 A cm^{-2} discharged conditions also revealed that hydrogen and carbon monoxide decreased as the current density increases, while methane did not decrease. These results suggest that hydrogen and carbon monoxide were effectively used as fuel species for eletrochemical oxidation reactions, while methane was not used.

These results of gas analysis reveal that methane was effectively converted to hydrogen and carbon monoxide in the radial direction on practical size Ni-SDC anode, and that hydrogen and carbon monoxide were effectively used as fuel species for eletrochemical oxidation reactions.

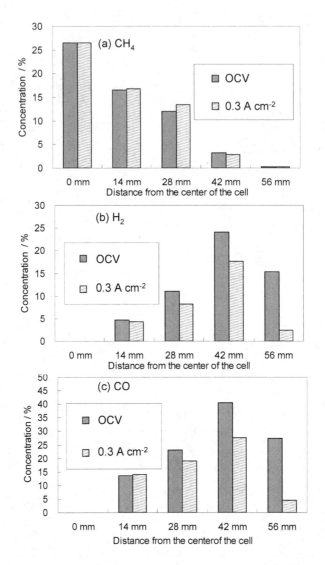

Figure 11. Gas analyzing results for (a) CH₄, (b) H₂, and (c) CO in radial direction for a single cell by feeding carbon dioxide-reformed methane (CH_4 /CO_2 = 1/3) with the conditions of open circuit voltage and 0.3 A cm^{-2} at 750°C after 2000 h operation.

CONCLUSIONS

Temperature dependence of the carbon dioxide reforming of methane for SOFC power generation revealed that the direct internal reforming of methane attained sufficient level of conversion for power generation with methane at 750℃, while the direct internal reforming gave insufficient conversion at 700 and 650℃.

Dependence of fuel conditions of the carbon dioxide reforming of methane for SOFC power generation also revealed that methane fuel was sufficiently reformed on Ni-SDC anode for the direct internal carbon dioxide reforming regardless of operation runs for CH_4 / CO_2 ratios being 1/4, 1/3, 1/2 at 750℃.

The comparison between the steam reforming and the carbon dioxide reforming also reveals that the cell performance with the carbon dioxide reforming at high current densities was higher than that for the steam reforming.

The results of gas analysis reveal that methane was effectively converted to hydrogen and carbon monoxide in the radial direction on practical size Ni-SDC anode, and that hydrogen and carbon monoxide were effectively used as fuel species for eletrochemical oxidation reactions.

It is concluded that carbon dioxide reforming attained sufficient level of conversion for power generation with CH_4 / CO_2 ratios being 1/4, 1/3, 1/2 at 750℃ on Ni-ceria-based oxide cermet anode of practical size SOFCs.

REFERENCES

[1]A. Atkinson, S. Barnett, R. J. Gorte, J. T. S. Irvine, A. J. Mcevoy, M. Mogensen, S. C. Singhal, and J. Vohs, Advanced anodes for high-temperature fuel cells, *Nature Materials*, **3**, 17-27 (2004).

[2]A.L. Dicks, Advances in catalysts for internal reforming in high temperature fuel cells, *J. Power sources*, **71**, 111-122 (1998).

[3]E. P. Murray, T. Asai, and S. A. Barnett, A direct-methane fuel cell with a ceria-based anode, *Nature*, **400**, 649-51 (1999).

[4]H. Kishimoto, T. Horita, K. Yamaji, Y. Xiong, N. Sakai, M. E. Brito, and H. Yokokawa, Feasibility of n-Dodecane Fuel for Solid Oxide Fuel Cell with Ni-ScSZ Anode, *J. Electrochem. Soc.*, **152**(3), A532-8 (2005).

[5]H. Kishimoto, K. Yamaji, T. Horita, Y. Xiong, N. Sakai, M. E. Brito, and H. Yokokawa, Reaction Process in the Ni-ScSZ Anode for Hydrocarbon Fueled SOFCs, *J. Electrochem. Soc.*, **153**(6), A982-8 (2006).

[6]I. Gavrielatos and S. Neophytides, High Tolerant to Carbon Deposition Ni-based Electrodes under Internal Steam Reforming Conditions, *ECS Trans.*, **7**(1), 1483-90 (2007).

[7]N. Laosiripojana, S. Assabumrungrat, and S. Charojrochkul, Steam Reforming of Ethanol over Ni on High Surface Area Ceria Support: Influence of Redox Properties on Catalyst Stability and Product Selectivities, *ECS Trans.*, **7**(1), 1717 (2007).

[8]J. V. herle, Y. Membrez, and O. Bucheli, Biogas as a fuel source for SOFC co-generators, *J. Power sources*, **127**, 300-12 (2004).

[9]J. Huang, and R. J. Crookes, Assessment of simulated biogas as a fuel for the spark ignition engine, *Fuel*, **77**, 1793-1801 (1998).

[10]M. Kawano, M. Matsui, R. Kikuchi, H. Yoshida, T. Inagaki, and K. Eguchi, Steam reforming on Ni-samaria-doped ceria cermet anode for practical size solid oxide fuel cell at intermediate temperatures, *J. Power Sources*, **182**, 496-502 (2008).

[11]M. Kawano, K. Hashino, H. Yoshida, H. Ijichi, S. Takahashi, S. Suda, and T. Inagaki, Synthesis and characterizations of composite particles for solid oxide fuel cell anodes by spray pyrolysis and intermediate temperature cell performance, *J. Power Sources*, **152**, 196-9 (2005).

[12]M. Kawano, H. Yoshida, K. Hashino, H. Ijichi, S. Suda, K. Kawahara, and T. Inagaki, Studies on

synthetic conditions of spray pyrolysis by acids addition for development of highly active Ni-SDC cermet anode, *Solid State Ionics*, **177**, 3315-21 (2006).

[13]M. Kawano, T. Matsui, R. Kikuchi, H. Yoshida, T. Inagaki, and K. Eguchi, Direct Internal Steam Reforming at SOFC Anodes Composed of NiO-SDC Composite Particles, *J. Electrochem. Soc.*, **154**(5), B460-65 (2007).

[14]T. Ishihara, M. Honda, T. Shibayama, H. Minami, H. Nishiguchi, and Y. Takita, Intermediate Temperature Solid Oxide Fuel Cells Using a New LaGaO$_3$ Based Oxide Ion Conductor -I. Doped SmCoO$_3$ as a New Cathode Material-, *J. Electrochem. Soc.*, **145**, 3177-83 (1998).

Author Index